D0240434

SPECIAL MESSAGE TO READERS

This book is published under the auspices of

THE ULVERSCROFT FOUNDATION

(registered charity No. 264873 UK)

Established in 1972 to provide funds for research, diagnosis and treatment of eye diseases. Examples of contributions made are: —

A Children's Assessment Unit at Moorfield's Hospital, London.

•

Twin operating theatres at the Western Ophthalmic Hospital, London.

•

A Chair of Ophthalmology at the Royal Australian College of Ophthalmologists.

•

The Ulverscroft Children's Eye Unit at the Great Ormond Street Hospital For Sick Children, London.

You can help further the work of the Foundation by making a donation or leaving a legacy. Every contribution, no matter how small, is received with gratitude. Please write for details to:

THE ULVERSCROFT FOUNDATION,
The Green, Bradgate Road, Anstey,
Leicester LE7 7FU, England.
Telephone: (0116) 236 4325

In Australia write to:
THE ULVERSCROFT FOUNDATION,
c/o The Royal Australian and New Zealand
College of Ophthalmologists,
94-98 Chalmers Street, Surry Hills,
N.S.W. 2010, Australia

Vince Smith holds an MSc in Behavioural Ecology and has devoted much of his life to wildlife conservation. From 1987 to 1993 he was senior keeper of apes at Chester Zoo, also working as a consultant for the Jane Goodall Institute in the Republic of Congo. In 1993 he left England to set up and manage the Sweetwaters Chimpanzee Sanctuary in Nanyuki, Kenya. Since 2000 he has been living in Rwanda, where he manages the Dian Fossey Gorilla Fund's mountain gorilla conservation programme for Rwanda, Uganda and Congo (DRC). He is married with one son, Oliver.

SOPHIE'S STORY

In 1990, while Vince Smith was working as a senior keeper at Chester Zoo, a newborn chimpanzee in his care was abandoned by her mother. Named Sophie, the baby chimp was taken home and hand-reared by Vince and his wife. Six months later another new baby arrived: Oliver, their son. This is an enthralling account of Sophie's life: her early years in the English countryside with Vince and his family; her traumatic removal from the family she adored into a captive zoo world; her repatriation to Africa and eventual reunion with her human foster family; and, finally, her integration into a semi-wild group of chimpanzees.

VINCE SMITH

SOPHIE'S STORY

Raising a chimp in the family

Complete and Unabridged

CHARNWOOD
Leicester

First published in Great Britain in 2003 by
Portrait, an imprint of
Judy Piatkus (Publishers) Limited
London

First Charnwood Edition
published 2004
by arrangement with
Judy Piatkus (Publishers) Limited
London

British Library CIP Data

Smith, Vince, *1959* –
Sophie's story: raising a chimp in the family.—
Large print ed.—
Charnwood library series
1. Smith, Vince, *1959* – —Family
2. Foster care of animals 3. Chimpanzees
4. Human-animal relationships
5. Large type books
I. Title
599.8′85

ISBN 1–84395–529–6

Published by
F. A. Thorpe (Publishing)
Anstey, Leicestershire

Set by Words & Graphics Ltd.
Anstey, Leicestershire
Printed and bound in Great Britain by
T. J. International Ltd., Padstow, Cornwall

This book is printed on acid-free paper

Picture Credits:
All photographs by the author except the following:

Russell Clarke: page 5 *bottom*
Debby Cox: page 7 *bottom*
Tim Hoolahan: page 1 *bottom*, page 3 *top*, page 7 *middle*
Audrey Smith: page 1 *top*
Photographer unknown: page 4 *top*, page 8 *top*

Every effort has been made to identify and acknowledge the photographers whose pictures appear in this book. Any errors or omissions will be rectified in future editions.

To my Oli.
Love you always,
Dad

Contents

Part Four: Becoming Chimp

Acknowledgements

I would like to thank my literary agent, Patrick Walsh of Conville and Walsh, for all his endless encouragement. I would also like to thank Peter Tallack and Richard Dawes for their help and advice in editing this book, and Penny Phillips of Piatkus Books for her backing and support. I am also grateful to Lauren Harris for sparing the time to share some of his immense knowledge on left-side cradling, and to Tim Hoolahan for allowing me to use some of his photos of Sophie in this book. And, of course, special thanks to my family, Audrey and Oliver, for following me on my African adventure.

Preface

As the blur of tiny acacias alongside the runway gradually slowed into focus, I relaxed back into my seat not knowing quite what was in store for me. My eight-hour flight to Nairobi had provided me with ample time to take stock of the events that were rapidly unfolding in my life. I was embarking on a new adventure, a new beginning. I had accepted a position, offered to me by the renowned primatologist Dr Jane Goodall, to manage a chimpanzee sanctuary in Kenya. Jane had already had an early influence on my life. As a child I had been awestruck by her television documentaries on the chimpanzees of the Gombe Stream Reserve in Tanzania, and had been particularly moved by the story of an infant called Flint, who refused to 'grow up' and resisted his ageing mother's attempts to wean him — an episode that ended tragically with the death of both of them. From that moment on, I was captivated by primates, though never in my wildest dreams did I imagine that I would be working for this remarkable woman in Africa. Apes possess such wonderful, endearing qualities that they can change your life, and soon, as has happened to many others before me, both apes and Africa were in my blood.

The opportunity to work in Kenya came as rather a surprise. I had previously been working for Jane as a consultant in Brazzaville, in the Republic of Congo, and for some time had believed that I would shortly be working there full-time. I loved the rainforest, and being in the Congo was like going back in time to a place full of romance and charm. But African politics soon destroyed that image, transforming the country into one riddled with hostility and insecurity. Following my consultancy, Jane wanted me to manage her chimpanzee sanctuary in Pointe-Noire, the Congo's second-largest city, near the coast. However, owing to the political turmoil, I was concerned for the safety of my family. Jane respected my concerns and offered me the alternative of managing another chimpanzee sanctuary in Kenya. I accepted the opportunity because it provided a healthier environment for my two-year-old son to grow up in. But there was another reason why I chose Kenya: at the back of my mind was the faint hope that this job might just give me what I had been praying for: the chance to free my daughter, Sophie, from a living nightmare. A bit dramatic, perhaps, but not when put into the context of the past few years of my life. Although my wife and son would shortly join me in Kenya, there was no guarantee that my daughter would be able to follow. You see, there was a minor problem. Sophie wasn't human — she was a chimpanzee.

Sophie was born into this world a captive. She was abandoned by her mother at birth, and I was given the chance to raise her. As a baby, she was just like any other human child: good-natured, mischievous, demanding of attention and always wanting to learn more about her surroundings. As far as she was concerned, I was her parent; she knew no different, and we quickly bonded, as in any other parent-offspring relationship. Although we were different species, I bonded with her as one would with a human baby one was fostering.

When my son, Oliver, was born, Sophie was just six months old, but much more advanced physically than a human child of equivalent age. The two of them lived seemingly parallel lives, and it was fascinating to watch them growing up together. As with all young children, there was no barrier between them in those early days because neither could speak and both ran around on all fours. Oliver had a unique and privileged upbringing as a child. He not only had as a friend a chimpanzee, who regarded him as a sibling, but he later grew up alongside other chimpanzees among the magnificent wildlife of Africa. Although they were more or less of equal size throughout the first seven years of their lives, Sophie always knew that Oliver was younger and more vulnerable and constantly watched over him.

Sophie and Oliver had one remarkable quality in common: they shared a parent. I was Oliver's father, but I was also a mother to Sophie. I use the term 'mother' carefully, for chimpanzees do not have paternal relationships. As a female chimpanzee may be mated by several males, the identity of the baby's father remains a mystery to the mother, the infant and all the males in the community. I was, of course, unable to sit Sophie down on my knee and explain to her, 'Look, Sophie, there's something I need to tell you. I'm not your real mother. I'm not even female. In fact we're not even the same species.' But despite this common bond between Oliver and Sophie, their parity ended there. Oliver had two parents who loved him deeply and were always there whenever needed. Sophie, by contrast, never had the luxury of having a parent constantly at her side to protect and support her in times of strife and to correct her about what was right and wrong. Of course, we spoilt her and gave her all the affection possible, more than she could ever get from a chimpanzee mother, but there were always limitations.

As she grew older and larger, so areas of the house became out of bounds until she eventually had to sleep on her own at night. Just like young Flint in Jane's film, she always resisted this, and it must have seemed incomprehensible and incredibly cruel to her. She was, of course, not human and yet, through no fault of her own, had been born into a human world. In this world our relationship had to be finite, and I watched this relationship slowly slipping away, like the sand in an hourglass, leaving me with a tremendous sense of guilt, which weighed heavily on me.

So what is it about apes that so fascinates and captures the human imagination? Apes have been studied for many years and have always evoked strong feelings in people. They display remarkable similarities to humans in their social behaviours as well as in their ability to communicate and to use tools and even medicinal plants. On the surface, a chimpanzee, walking on four legs, with its hirsute body and protruding face, appears just like another species of monkey. But appearances can be misleading. Chimpanzees are not monkeys; they are apes, along with gorillas, bonobos (or pygmy chimpanzees), orang-utans and gibbons. Furthermore, apes and humans are classified as members of the same superfamily, Hominoidea. Recent scientific evidence indicates that our earliest ancestors, the australopithecines, were, in their body proportions, more akin to modern-day bonobos than to modern-day humans. As a result, chimpanzees provide perfect models for exploring such far-reaching questions as, what makes us human and where do we come from?

Molecular biological research has compared deoxyribonucleic acid (better known as DNA) to establish genetic distances between the ape species and approximate their dates of evolutionary divergence. The findings suggest that the ancestors of orang-utans branched off from the ancestors of gorillas, chimpanzees and humans some 12 to 16 million years ago, with the ancestors of gorillas branching off from those of the chimpanzees and humans some six to eight millions years ago, and the ancestors of chimpanzees and humans then diverging some four to six million years ago. So humans evolved not from chimpanzees, but from a common ancestor that is now extinct, which in turn evolved not from monkeys, but from an even older ancestor that is also extinct.

Chimpanzees are our closest living relatives. But, more interestingly, we are also theirs. In fact the two species of chimpanzee — the common chimpanzee and the bonobo — are both 98.4 per cent genetically identical to us. Indeed they are more closely related to humans than they are to gorillas, and much more so than they are to orang-utans. Gorillas are 97.7 per cent genetically similar to humans and to both species of chimpanzee, while orang-utans are a mere 96.5 per cent similar to us and chimps.

Chimpanzee and human blood is very hard to distinguish. Our haemoglobin — the oxygen-carrying protein that gives blood its red colour — is identical to chimpanzee haemoglobin. We are in essence blood relatives. When it comes to their genetic closeness to monkeys, chimpanzees — like humans — are only distantly related. To put this into perspective, chimpanzees and humans are genetically more closely related to each other than is the horse to the zebra. Being genetically so similar, humans tend to relate more closely to chimpanzees than to any other species of animal. Because chimps also share so many of our behavioural qualities, both good (sharing and caring) and bad (aggression and jealousy), they are used as models for the interpretation of some of our human characteristics, and for research into the origins of these, as well as to establish our place in nature, both present and past. Chimpanzees can live to 50 years or more. But, sadly, if present trends continue, all the apes throughout the world may be extinct from the wild within a single chimpanzee's lifespan.

Once you look at a chimpanzee from this viewpoint, a different idea of these animals begins to develop. I hope that some of Sophie's experiences described in this book will help to flesh out this picture. For, beneath the surface, beneath that thick concealing mat of body hair, chimps have feelings and personality — they have a soul. Is the possession of emotions an attribute unique to humans? I don't believe so. Affection and other emotions are, of course, subjective, but it would be chauvinistic to believe we are the only ones with the capacity for conscious thought, creativity, sensitivity and culture.

But it would be equally mistaken to suggest that humans and apes are intellectually comparable. Most definitely we are not. Humans, with their intellectual capacity, language and culture, are clearly on a different level from the rest of the animal kingdom. But then, conceivably, so too are the great apes, although they are on a much lower tier than humans. Maybe the immense gap that we perceive to exist between all other animals and ourselves is perhaps not quite as wide as we would like to believe. And, for some, the similarities between humans and apes are too close for comfort.

This is the biography not of a human being but of a being nevertheless. Sophie's non-humanness should not detract from the significance of her existence. In this book, while trying to avoid being unnecessarily anthropomorphic, I have attempted to unravel the experiences of her fascinating life as I believe she witnessed them on her long and eventful journey from England back to Africa.

Part One

The Wilds of Shropshire

1

A Dry Run

In the summer of 1984 my wife, Audrey, and I took the bold decision to move away from the hustle and bustle of London life to find our dream home in the country. We had been working for the past five years with British Telecom, and I had grown dissatisfied with my nine-to-five life of complacency in the big city. At the age of just 25, I felt I needed a more stimulating challenge.

After an extensive search across the English countryside, encompassing seemingly everywhere from Cornwall to Lincolnshire, we stumbled on what we believed was the perfect place — a small, detached cottage with 16½ acres (6.7 hectares) of land, perched high on a secluded hill in the remote village of Meadowtown, roughly halfway between the medieval market towns of Shrewsbury and Ludlow. So, after convincing ourselves that we were doing the right thing, we took a gamble — one of many in our lives — and both resigned from the security of our jobs.

The views from our house were breathtaking and, with unashamed bias, we always maintained that there was no more glorious a spectacle in the entire country. From the house we looked down across a steep-sided valley to far-off hills that formed part of the border between England and Wales. The criss-cross patterns of distant fields together made up a patchwork quilt of staggering beauty and widely varying colour, with various hues of pasture green interspersed with the brown of freshly ploughed earth and the stunning yellow of rapeseed. There were no villages to be seen on this canvas, no major roads, no pylons — just mile upon mile of unspoilt countryside. It was a setting to inspire any artist or poet.

Although the surrounding countryside seemed perfect, our home was anything but. It was extremely basic, consisting of just two bedrooms, a kitchen-dining room and a lounge. The lounge, at least, was attractive, with whitewashed stone walls, exposed oak beams on the ceiling, flagstones on the floor and a large open fireplace. The house was more than 200 years old and was originally a miner's cottage. In those bygone days a person would only have to erect a chimney stack to claim the land. There was no inside toilet and, worse still, no mains electricity. Instead we made do with a small Lister diesel generator. From the house to the road far below ran a long, winding and bumpy track, and we often joked to our visiting friends as they were leaving, reminding them to pick up their car exhaust on the way down.

To put it mildly, the house left much to be desired and there was a lot of work to be done. But we loved it all the same and were determined to make a life for ourselves there. We were roughing it, but we were young and very much in love, so missing some of life's luxuries seemed a sacrifice worth making. More importantly, we could see that the house had enormous potential, especially with all its land, which was the main reason we had fallen in love with it in the first place. Much of the land was hilly and wild and covered by gorse. To us the acreage seemed much larger than the figure on paper, and if we could have ironed out all the hills and creases then it would have been even bigger! As it was, it took at least 25 minutes to walk all around the perimeter.

We had two ponds, one large one in front of the house and one small in the middle of our land, both fed by natural springs. The large pond had a small island in the middle and was stocked with brown trout, mirror carp, rudd and sunfish. The small pond was stocked with minnows. The pond in front of the house was surrounded by an area of marshland, where among the reeds grew beautiful marsh orchids and marsh marigolds. Sometimes we rowed around this pond in a small boat or just sat on our island for hours and watched the fish — it was very relaxing. And if all that wasn't enough, then at the bottom of our land was a small wood of mature deciduous trees, complete with a narrow stream running through the middle. All in all, it was our little piece of paradise.

Of course, all this land needed to be managed. We couldn't just let the grass grow wild, as it would soon become matted and tough. So we bought two young heifers — a pure Jersey and a

Guernsey-Charolais cross — and a small flock of eight Manx Loughton sheep, a rare breed. From an early age I had had an interest in marsupials and one of my ambitions was to open a small zoo specialising in Australian animals. So I used deer fencing to form a paddock of one and a half acres (0.6 hectares) and bought a pair of Bennett's wallabies.

Audrey was the first to find work. A qualified seamstress, she had worked in London's Savile Row, famous for its haute couture, and so decided to start her own small business at home. She also harboured ambitions of one day entering the world of fashion design. She soon became well known in the locality and it wasn't long before she was turning away requests to make or alter dresses, men's suits, children's clothes and upholstery.

Finding a job that interested me, however, proved more difficult than I had initially imagined. But first I had to learn to drive. Fortunately, I passed my driving test quite quickly and we celebrated by immediately buying our first car — an old white Land Rover. At last we had wheels, so no more walking two miles home from the nearest bus stop laden with heavy shopping. The car also allowed me to look for employment further afield. Because of my wallabies, I already knew the management of Chester Zoo, so in 1987, a couple of months short of my twenty-eighth birthday, I took a job working with their primates. Chester had (and still has) a large collection of apes, monkeys and lemurs. But it was especially proud of its group of 25 chimpanzees, which lived on a one-acre (0.4 hectare) island and were the second-largest group in Europe. The money wasn't good but it paid the mortgage and gave me some form of job satisfaction, as I was now involved in conservation work. However, because the journey to work took an hour and a half, I was not only having to get up very early each morning but also spending most of my wages on fuel. Working with monkeys, I was earning peanuts!

Winter was especially hard. I would get up at the crack of dawn, scrape the ice off the windscreen, pray the car would start and then drive off in the dark with my hands frozen to the steering wheel. Audrey would spend the day alone in the house until I returned home in the dark around seven in the evening. The demands of our animals and the sewing jobs she was getting meant she didn't have time to be bored or lonely. But, whenever she could, she strolled down our hill and popped in to one of our neighbours for a chat and a cup of tea. She was very popular in the village. It was a hard, harsh life in those early days, but we kept telling ourselves that it would all be worth it in the end.

All my life I had been deeply interested in wildlife and conservation — ever since, as an eight-year-old, I had won my first goldfish at a fairground at the Epsom Races. To this day I still remember how I eagerly carried my prize home in a plastic bag, making sure not to shake the water. I remember gently placing the fish in its new home, a small, round fish bowl that my mum had generously bought me at the fair, and watching proudly, transfixed, as my first-ever pet swam clockwise round and round the crystal waters of its pristine world. And I still remember the feeling of utter dismay when Goldy decided to be adventurous for a change and swim anticlockwise, revealing to me its hidden secret, an eyeless profile — the right one as I recall.

My father could see that I was keen on animals and on my ninth birthday bought me a 20-volume set of animal encyclopaedias. I was fascinated by these and spent hours educating myself about everything from aardvarks to zebras, carefully turning the pages so as not to crease them. But more than any other species, it was marsupials that captured my imagination. As I grew older, so my collection of specialist wildlife books grew with me.

Our first winter in Shropshire was the worst, and the snows fell long and deep. The winds blew much of the snow from our neighbour's field on to our track, where it collected until it lay several feet deep. We ended up snowed in for two weeks. It was particularly depressing to see that the west side of the valley had at least two hours' more sunshine each day than we did. And to make matters worse, as a result of the extra sunshine the snows on that side had thawed after just five days. Audrey and I took it in turn to get up in the middle of the night and, armed with a torch, stumble across the fields to check on the animals. Our life was far removed from our cosy existence in London, but in some ways it was quite exhilarating. Gone was our central heating and all the other comforts. Instead we would cuddle up in the evenings in front of the television, and warm our toes in front of a portable Calor-gas heater.

Living so close to nature made us very aware of the changing seasons, and we would yearn for springtime. When it finally came it was a huge relief to animals and humans alike, and we felt instantly rejuvenated. The hills, which over the winter had become faded, were now vivid green with lush growth. The gorse was rapidly transformed from an uninteresting prickly bush to an attractive, yellow-flowering shrub. Wild flowers such as foxgloves, campions and poppies grew in abundance among the gorse, all blossoming in harmony. The gorse provided excellent cover for the many rabbits that chose to share our land, and we'd sit on the grass in the warm sunshine and watch the young bucks sparring over their prospective mates. The gorse was also excellent cover for our cat, Sabby, who enjoyed watching the rabbits for entirely different reasons.

We didn't want to go through too many winters like the first one again and were determined to make home improvements during the summer. So we obtained permission from our local council to build an extension to the cottage. We added two bedrooms, a large kitchen and a bathroom with a toilet. Then we paid for mains electricity to be connected to the house. We spent a lot of money but every penny was worth it, particularly as the work greatly increased the value of our property.

All around us there seemed to be a frenzy of

wildlife activity. Treecreepers, nuthatches and both greater and lesser spotted woodpeckers regularly visited the damson trees in front of the house and we soon developed a keen interest in ornithology. We were particularly fond of our resident breeding pair of redstarts, listening out for their distinctive calls to reassure ourselves that they had survived another season. Sabby also liked listening to the redstarts and we were dismayed one morning when she entered the house carrying the female in her mouth and proudly paraded its body across the kitchen floor in front of us. It sang no more! Other than the redstarts, the rest of the animals seemed to be breeding everywhere and nowhere more so than in our pond. The fish were prolific, especially the sunfish, which would vigilantly guard their fry as they herded them about in a tight-knit school. Frogs appeared in abundance, and it wasn't long before much of the water was turned into a heaving, throbbing mass of spawn. Our sheep too were lambing, and soon our flock of eight had increased to 15. And to top it all, there was a distinct bulge kicking in the wallaby's pouch.

One day, when I was at the bottom of our land tending the sheep, I heard Sabby calling out from a distance. She was meowing in a rather unusual way. She saw me in the distance and anxiously rushed over to me. I was surprised that she should come looking for me, as, like most cats, she was very independent. At the time, Sabby was heavily pregnant after an encounter with an uncouth and disgusting stray cat. I knew instantly that she was close to giving birth and had travelled across several fields just to find me — it was like a scene from the film *Lassie*. Astonished and touched by her need for my reassuring presence at such an important moment in her life, I carried her into the house. No sooner had I placed her in her basket than she began to have contractions. Ten minutes later she had given birth to six kittens.

The animals weren't the only ones trying to have babies. We too had been trying for a child for almost two years, although only half-heartedly. It was not something we had been taking too seriously, but nevertheless we'd stopped taking precautions to see what would happen. Well, nothing did happen! Being paranoid, we soon became worried that maybe we would never be able to achieve what should be a simple process of nature.

I had first met Audrey in London, where we were both working as French linguists for British Telecom. She was Creole Mauritian and stunningly beautiful, and as soon as I first laid eyes on her I was in awe of her exotic looks. She was born in Beau Bassin, on the island of Mauritius. The youngest of four children, at the age of seven she moved to England with her family. She grew up in north London, and in some ways I always felt she was more typically British than me, especially with her roast beef and Yorkshire pudding on Sundays and the way she religiously watched the Queen's speech at Christmas — something I never did. My life soon became one of misery and frustration as I spent endless sleepless nights thinking about her.

But eventually I summoned up the courage to ask her out in my own awkward way and she agreed. I was pleasantly surprised to discover that she held similar feelings for me.

Meanwhile, at Chester Zoo, two of the chimpanzees — Mandy, a low-ranking, hand-reared female, and Halfpenny, a high-ranking, mother-reared female — were pregnant and due to give birth at any moment. The two chimps couldn't have been more different in personality. Halfpenny was powerful and elegant. She was always aloof with her human caretakers, as if they existed only on the periphery of her captive world and were there simply to feed or annoy her. Yet with the other chimpanzees she was full of confidence and assertiveness. Mandy, by contrast, was a timid and diminutive creature within her group, but loved interacting with humans. Whether or not it was because she posed no threat to the other chimps, she had the knack of avoiding trouble and seemed relatively popular. Halfpenny was pregnant for the second time. She already had a healthy infant called Sarah. Mandy had given birth on three occasions but each time had abandoned her baby. She had been hand-reared herself and so had not developed sufficient maternal skills of her own. Usually a female will produce a baby once every five years, but if she loses her baby, she is capable of giving birth again a year later.

As I drove home I thought about the chimps and wondered if this time Mandy was going to surprise us all by being a good mother. But my thoughts kept returning to Audrey and our predicament and I wondered if we would ever have children. I sensed too that she was becoming increasingly desperate, and in the darkness of the night her heavy swallowing of tears revealed her true state of emotion. I knew how much having a baby meant to her and equally knew what a wonderful mother she would make because she loved nothing more than to play with children. It would be a tragedy for someone like Audrey to be deprived of motherhood and I felt bad that we hadn't tried for a baby earlier, when we were younger, instead of just taking such things for granted.

On 20 September 1990 our alarm clock rang at six o'clock as usual. As I wrestled with sleep, I was surprised to see that Audrey was already up. I could hear her in the bathroom — only she wasn't singing or splashing, as she usually did, but making retching noises. This was now the third morning in a row that she'd woken feeling sick. Worried she might be ill, we decided it best that she visit the doctor in Shrewsbury. So early that morning I dropped her off at the surgery and continued on my way to Chester.

When I entered the chimp house, I found the alpha male, Boris, sitting by the bars waiting to greet me. Boris enjoyed interacting with the staff, and this morning he was in a particularly good mood, wanting me to chase him down the corridor. There were five males in the group and 20 females. So for Boris to maintain his alpha status was no mean feat. When playing with Boris, you had to know when it was time to stop, as there was a fine line between play and aggression. If he became overexcited, his play gradually descended into an aggressive display. As a rule, the chimps had to be treated with the utmost respect. As friendly as they could be, they were capable of turning aggressive at any given moment. It was generally agreed among the zoo staff that, along with the elephants, the chimps were the most dangerous animals to work with, because of their strength, agility and intelligence. Feeding the chimps was the most critical time, as they were fed manually through the bars, which entailed placing food directly into their hands. This ensured that each individual received his or her fair share. It also maintained contact with, and acceptance from, the chimps.

The first few months on the chimp section were always a dangerous time for any new keeper. And if a chimp had a gripe against you, then it might just decide to sacrifice the food being offered for the chance to bite the hand that was feeding it. Certain chimps liked to spit at the keepers; not to get your attention, but simply because they didn't like you. Usually, if the chimps wanted to get your attention, they blew raspberries. Some of the chimps had poor manners and would throw faeces at you, especially if you were new on the section or were unpopular. Boris and two females, Heidi and Rosie, were the worst for this.

When Steve Cook joined the chimp section for a few months, Rosie and Heidi gave him a torrid time. Each morning they would be waiting by the entrance for him to enter the building so that they could pelt him with faeces. One such morning Steve cautiously poked his head round the door. He asked me if the coast was clear, and I assured him it was. But, shortly afterwards, Heidi appeared with her hand hidden behind her back and began nonchalantly walking towards us. She looked decidedly suspicious, so I warned Steve that she might be up to no good. At the same time I took a couple of steps away from him — choosing self-preservation over valour. As we suspected, when Heidi was within a few feet of Steve, she suddenly revealed her hand, which was carrying a mountain of faeces. By this time it was too late for him to react, let alone escape, and Heidi's ammunition landed full in his face. These annoying interactions were regular occurrences. However, this particular incident was interesting because it demonstrated great intelligence on the part of Heidi, as it was clearly premeditated, employed deception and was cunningly executed.

It was seven o'clock that evening when I walked into the house and called out, 'Hello, I'm home.' Audrey was standing in the kitchen with her back to me. She turned around and was smiling, yet tears were running down her face.

'What did the doctor say?' I asked anxiously. Audrey gently put her arms around my waist and announced proudly that she had taken a pregnancy test and that the results were positive. We were at last destined to become parents. It turned out that all along she had been miscalculating the beginning of her fertile cycles, as she was counting from the end of her periods instead of the beginning, and as a result was mistiming the peak of ovulation. We stood there for a while hugging each other, overcome with emotion.

Five days later the birth of a small primate occurred — an event that was to shape and change the course of our lives for ever. The baby was born to Mandy, a good-looking chimp with delicate features set in a jet-black face. People might ask how you distinguish between a good-looking chimpanzee and an unattractive one. After all, don't they all look the same? Well, let me assure you that once your eyes have become attuned to their features, no two chimpanzees will ever look alike again. Mandy's baby, in my obviously biased opinion, was an extremely attractive chimp. The previous day Halfpenny had also given birth to a healthy female, Kaylie. As time passed, Kaylie provided a useful yardstick with which to gauge Mandy's baby's development.

Halfpenny proved an excellent mother, being very protective and attentive. Mandy, conversely, again abandoned her baby on the cold concrete floor. The father of this baby was Boris. Good-looking and well proportioned, he was exceptionally intelligent, with a strong sense of mischief that bordered on humour. Boris was light-skinned, with long, powerful arms and legs. He had a distinctive walk, placing his two arms forward on to the ground and then swinging his legs and body between them — like walking on crutches.

Boris had been donated to the zoo by his American owner, Hester Mundis, when he became too large for her to handle. As coincidence would have it, she wrote a book about Boris, entitled *No, He's Not a Monkey, He's an Ape and He's My Son*. Mandy's baby inherited not only some of his facial characteristics but also his unusual walk. So, in a peculiar way, this baby was to follow in her father's footsteps. That afternoon we offered the baby back to Mandy in a last-ditch effort to persuade her to rear her infant but once more her helpless cries were ignored.

Mandy's rejection of her offspring presented the zoo with a problem: what do we do with this abandoned baby? The Curator of Mammals, Nick Ellerton, called me into his office and asked me if I would be willing to hand-rear the baby. I didn't need much persuading, but I first checked with Audrey to see if she had any objections. I rang her from Nick's office and she immediately said yes to the idea. Apart from anything else, we agreed the rejected chimp would act as a useful dry run to help us develop our own parental skills.

So, late that afternoon, I wrapped this tiny baby in blankets and placed her in a cardboard box with a warm hot-water bottle beneath the blankets and drove the long journey to her new home in Meadowtown. I carefully carried her into the house, where Audrey was eagerly awaiting the new addition to our household. She had recently resigned from her job in Ludlow, so she welcomed the company. It was also a great thrill for her to be able to experience for herself the privilege of working closely with a rare species like chimpanzees after having endured so many of my own tedious zoo stories.

When I first joined the zoo I had always hoped that one day I might get the chance to hand-rear some exotic animal, perhaps a lion or a tiger. Now, I had to pinch myself that this miniature, human-like body living in our house was really a chimpanzee. She was so small, weighing only two pounds thirteen ounces (1.3 kilograms), and although covered in thick, dark hair on her back, was completely hairless on her chest and arms. This lack of body hair gave her a distinctly human appearance.

Chimpanzee mother's milk is similar in composition to human milk. So we fed her on SMA human baby milk substitute — roughly half an ounce (15 ml) every two hours — and, because of her tiny size, had to use a small syringe. She didn't have a bowel movement for the first day. Initially she was only urinating in her nappy. Then, late the next morning, I changed her nappy and found her first stool. It was important that she excrete this sticky, greenish-black stool, known as the meconium, as it consists of, among other things, all the amniotic fluid, mucus and bile collected from the mother's uterus. I was a little concerned that she'd never had the opportunity of drinking her mother's colostrum. This substance, produced by the mother for a few days before milk comes, is rich in antibodies and therefore important for protecting the baby against infection from bacteria and viruses that her mother may have come into contact with and developed a resistance to. She would have to make do with just her powdered milk and I could only hope that this would not prevent her from developing into a strong, healthy individual.

After a week I decided to feed her on demand, letting her tell me when she was hungry and not the other way round. This was a much more natural way of feeding a baby. To begin with, she would drink between a quarter and three-quarters of an ounce (7-21 ml) every two to three hours, depending on how hungry she was. If she became constipated I would gently rub her tummy, which would stimulate her bowel movement. The cardboard box became her bed for the first couple of months of her life. At night I set my alarm clock to wake me every two hours so that I could feed her. It was exhausting at first, especially as I was spending three hours each day driving to and from work, and I began to have doubts whether I'd be able to cope, but my body and mind soon adapted.

We decided not to name her initially, but to wait until we felt sure she was going to survive. After about a week I came up with the name of Betty. I don't know why, but at the time I felt it suited her. However, when I checked the chimpanzee stud-book records, I saw there was already a chimp with that name. So I had to find another name unique to her. I had always liked the names Emma and Emily, probably because of the hit record by Hot Chocolate. But once again these two names were already spoken for. Eventually I settled on a name that no other captive chimpanzee had in Britain — Sophie.

After just ten days we had our first mini-crisis, when Sophie developed diarrhoea. To make matters worse, she stopped drinking, which resulting in her rapidly becoming dehydrated. Her eyes sank deep into their sockets and her body was covered in waves of creases. She became so weak she no longer had the strength to support her head. I feared the worst, wondering if she'd be alive the following morning. That night I laid her on my chest, hoping the sound of my heartbeat would have a soothing effect and give her the will to live. At first I was a bit nervous that I might roll over in my sleep and squash her, but my body somehow seemed aware of its surroundings even as I slept. Each hour throughout the night I forced her to take sips of rehydration fluid from a syringe. Every time I awoke, I anxiously checked on her, praying that she would still be alive, and was always relieved to find her still sleeping peacefully on my chest.

The following morning I could see a slight improvement. She had regained a little of her strength and was now able to hold her head up and look at me. More importantly, she had regained her appetite. Her eyes looked brighter and she eagerly gulped down her water. I could see that she was a fighter and was going to make it. My eyes, though, didn't look so bright. It was six in the morning, so I got dressed for my drive to Chester. I kissed Audrey goodbye and placed Sophie in the bed by her side. Audrey kept a non-stop vigil over her, making sure she received plenty of fluids. Sophie's diarrhoea stopped that same day, and just as rapidly as she had gone downhill, she returned to full health.

That night, Sophie again slept on my chest. Men are not designed for sleeping on. Unlike women, we are not endowed with two soft, padded pillows; instead we have a bony ribcage. As a result, Sophie invariably ended up having hiccups in the night. So I decided that sleeping like this wasn't a satisfactory solution, even though Sophie seemed quite content with her hard, lumpy mattress and no pillows. Instead we put her box in the bed next to us and I would place my arm over her until she fell asleep.

Because she was so small, I was unable to find any nappies to fit her. Even the smallest size was like a raincoat on her. So I improvised by making nappies out of small plastic freezer bags. I cut off the two bottom corners for her legs and then padded the inside with cotton wool. This proved a perfect fit and worked a treat for the first few months. Other than eating, her favourite pastime was filling these nappies, and I would normally change her several times a day and at least once during the night. I was always amazed that so much could come out of someone so little. I was doing this task so often that in the end I found the smell of her full nappy worryingly pleasant. Each morning after changing her, I would rub Johnson's baby oil into her body to prevent her skin from becoming too dry.

It was then time for her breakfast: SMA milk again. This was a complete meal in itself and was all she needed for the first few weeks of her life. As time went on, I began gradually introducing morsels of soft food to her diet, such as banana and Farley's rusks, which she adored, and so did I with my cup of tea. After each meal I would gently rub and pat her back until she let out a loud burp or two.

For the first few weeks Sophie spent much of her time lying in her box. She would stare up at the ceiling, kicking her legs in the air, and if she felt she was being ignored she would call out in her high-pitched, chattering squeak until she had regained our attention. Effectively having two pairs of hands, she would try to grab passing objects with either hand or foot. Humans and chimpanzees are both capable of precision grip using their thumb, which is opposable to their fingers — in other words, it is opposite the fingers instead of beside them. Humans have a more developed precision grip than chimpanzees, mainly because the chimpanzee thumb is shorter and so touches the sides of the fingers rather than the tips as in humans. However, chimps make up for this by being capable of precision grip with their toes as well. Unlike humans, chimpanzees, being semi-arboreal (equally comfortable in trees or on land), are blessed with an opposable big toe that acts like a thumb, allowing them to grasp branches and objects with their feet. I always felt it a shame that, as we humans evolved to become bipedal — walking on two feet — and therefore more flat-footed, we lost this useful appendage.

Apart from her big toe, everything else about Sophie's feet seemed very human, especially their cheesy smell. I was surprised at how powerful her grip was as a baby, and whenever changing her nappy I needed to keep my nose well out of reach of hand or foot. Despite her obvious strength, Sophie rarely held on to me; I had to support her weight at all times. If I relaxed my hold on her, she would slip down my body. At birth, mother-reared infant chimps are unable to cling without support. Normally they hold on to their mother's hair — usually beneath the protective area of her belly — while being cradled by her arm. The mother then moves around awkwardly with one arm supporting her baby. Human infants, by contrast, are completely helpless at birth. Being bipedal, a human mother can offer more attention to her infant when it comes to cradling, as her arms are free and not used for walking on all fours along the ground or climbing trees. But I had the impression that Sophie seemed to lag a little behind Kaylie in her physical development. I suppose I was probably over-supporting her and not encouraging her to use her own limb muscles. Maybe she felt confident in my secure grasp or maybe she was just plain lazy.

I was still enjoying my time at the zoo, even though I was spending 24 hours out of every eight days commuting along the same mundane roads. I would get to work in the morning tired and return home the same. But eventually it paid off, for it was there that I first met Jane Goodall. Jane had spent most of her life observing chimpanzees in the wild and had had tremendous success with books such as *In the Shadow of Man* and *Through a Window*. It was Jane who gave me my first opportunity to work in Africa. She was no longer continuing her own research there, but had set up a charity, the Jane Goodall Institute, which supports the ongoing research at the Gombe Stream Reserve, Tanzania, and also funds several sanctuaries for orphan chimps in Africa. She was giving a lecture at Chester on the wild chimpanzees of Gombe and naturally wanted to visit the zoo's chimps. Jane had an instant rapport with them, which was fascinating to watch. Normally the Chester chimps were quite aggressive towards strangers, but they treated Jane like a long-lost friend even though they'd never met before. At the time I was carrying Sophie, and so Jane spent a short while with me getting to know this baby chimp who was trying to sleep under my jumper.

Now that Sophie was also taking solid food, she seemed to grow faster than ever. Soon she was drinking out of a baby's bottle, which I rinsed in sterilising fluid after each feed. When she was about three weeks old I began to take her with me to work. The rhythm and rocking of the car put her to sleep within minutes, and she would usually lie on my lap or in a box and not wake up until the journey was over. It was very handy that she couldn't crawl yet, as I could leave her sleeping in her box, secure in the knowledge that she wouldn't be able to get up to any mischief.

At work I was spending most of the day looking after the orangutans — five Bornean and five Sumatran. I would place Sophie in the kitchen area of the orang-utan house and get on with cleaning and feeding them. Orangs are exceptionally intelligent and I believe the first film in the *Planet of the Apes* series got it right by portraying them as scholars. They are real thinkers and will spend hours trying to solve a puzzle. Chimps, by contrast, have a low attention threshold and will quickly lose interest or become frustrated. I have often been asked which species of ape I think is the most intelligent. Well, in a way this is an unfair question. Each species is adapted to its own set of environmental conditions: orang-utans are solitary, arboreal and mainly frugivorous, or fruit-eating; gorillas gregarious, mainly terrestrial and herbivorous; and chimpanzees gregarious, semi-arboreal, and mainly frugivorous. Generally, fruit-eating animals need a greater awareness of their environment than do herbivores, which more or less eat the vegetation they are sitting on. They require a mental map of their territory to know where fruiting trees are within their range and roughly when they will bear ripe fruit. Measuring each species against the other for cognition is a bit like taking a linguist, a civil engineer and a chemist and giving them a maths test — it's meaningless.

However, if pushed, I would say that the one individual that never ceased to amaze me with its intelligence was a male Bornean orang-utan named Anak. He was extremely possessive of his females, especially one he particularly liked, called Martha. However, he treated them like sex slaves and would drag them around by their hair wherever he went, so as not to lose sight of them.

He was also possessive about their food and, after gobbling down his, would steal theirs. Although he never physically harmed them, he made their lives a misery. Orangs in the wild are naturally solitary, with a male occupying a range which overlaps that of several females, who each have their own range. We would place a female in with him for mating and then try to separate them a few days later — mainly for the female's peace of mind but also because, over a long period of time, she would be unable to feed adequately. Trying to outwit Anak, so that we could rescue his female, was like playing a game of chess — a game he was very good at and would invariably win. He always knew what we were up to and was always one step ahead of the game. Nothing we did seemed to fool Anak, so it was very frustrating. Sometimes it could take over a week to separate them.

The orangs had access to an outside island and also an inside yard with a climbing frame. Two sliding doors, on either side of the yard and about 50 feet apart, gave access to four beds under cover that were interconnecting but could be separated. Here the orangs would sleep for the night. We would place food inside the beds and try to get Anak or his female to enter. If we opened just one door, then Anak would sit in the doorway so that we couldn't further open or close it. The only way to trick him was to open both doors. This would leave him trying to get the food while keeping one eye on the human staff, who were trying to close the door behind him, and the other eye on the whereabouts of his female, who was trying to get away from him through the other door. Anak's appetite would normally get the better of him, and he would rush inside, grab a handful of food and then tear back outside as fast as he could to see where the female was. If she was still on the climbing frame he would sit in the doorway and eat his food. But if she had ventured to the ground, he would drag her as far away from the door as possible.

Martha never enjoyed her time with Anak. She didn't take kindly to being pushed around and bullied or having her food stolen. She knew what we were trying to do and would do her utmost to play her part in escaping. But if she realised that she stood little chance of succeeding, then she wouldn't attempt a dash for the door.

There was one time when she had been stuck with Anak for about two weeks. Anak had become complacent and had let his guard down. He greedily ventured into the second bed to get some food. Martha saw her chance and made a frantic dash for the sliding door, which I managed to close seconds before he got there himself. Checkmate! The human-orang teamwork between me and Martha had finally paid off. Martha whimpered with relief as she sat down on her bedding and grabbed some bananas. Anak's reaction seemed so human. He shook his head and winced in disbelief and utter disappointment. He seemed truly annoyed with himself that he had fallen for that same old trick. Then he threw a tantrum, roaring and thumping the walls and bars of his cage with his fist. Martha finished her food and then curled up in her bedding and slept peacefully.

Like the chimps, some of the orang-utans would spit at the keepers. Two Bornean orangs in particular, Sibu and Lola, would take great delight in doing this. They would aim with great precision at the eyes or face. It wasn't so much that they didn't like us — Sibu was very friendly — it was more that they knew that this would attract our attention, as they had an endless craving for food, especially dried fruit and nuts. It was incredibly annoying and sooner or later we would give in. It was their form of bribery — give us some food or else! Walking down the corridor of the orang-utan house became like running the gauntlet through a shower of spittle. I would keep my lips turned in and pressed firmly together as I went about my work, stopping occasionally at the sink to rinse the foul-smelling saliva off my face. Sibu had developed this skill to a fine art, and whether you were standing still, walking or running, he would invariably hit his target with uncanny accuracy and from an impressive distance. He would adjust his range to match your speed and aim his fire just the required distance in front of you so that, as you continued on your way, the spittle would find its mark.

On one occasion when I was shouting at Sibu to cease fire, he hit the jackpot, his spit landing right in my open mouth. My face turned green and for the next half-hour I remained under the tap as I rinsed my mouth and gargled a hundred times over. Not long afterwards we decided to keep Listerine mouthwash in the kitchen for that very purpose. Sophie, meanwhile, was unaware of my ginger-haired tormentors, and spent most of her time sleeping peacefully in her box.

Back at Meadowtown, the animals were much more civilised and I would look forward to spending my days off with Audrey. Our collection of marsupials had grown significantly, and along with the Bennett's wallabies we now had wallaroos, dama and parma wallabies, long-nosed potoroos, grey short-tailed opossums and a pair of brush-tailed possums. I needed a Wild Animals Licence for the wallaroos, as these kangaroo-sized creatures are considered potentially dangerous. The wallabies took up a lot of our free time but gave us much pleasure. But our favourites were the potoroos, which are rabbit-sized wallabies. We had two adult pairs and successfully bred two babies. These were unbelievably cute, and when they emerged from their pouches they were no larger than a fat gerbil.

One weekend I was in the kitchen preparing the food for the wallabies when I heard a loud bump, followed by a brief silence and then, a few seconds later, frantic shrieking. I rushed upstairs to investigate and found Sophie whimpering on the floor. She had just learned how to crawl and had climbed out of her box and on to our bed. Like Christopher Columbus, she had wanted to explore the horizons of her new world — only to fall off the edge. She was unhurt of course, just a little startled that her first adventure should end so abruptly. But this meant that I could no longer leave her in her box. Now that she was mobile, she had more control over her own destiny and what she wanted to do. She was no longer a helpless soul, totally dependent on others. Now she could make decisions and do things for herself. Her first decision in life was that she was no longer satisfied with sleeping alone in her box. She wanted to sleep with me. The weather in the hills was becoming increasingly cool and I wasn't satisfied with using a hot-water bottle under her blankets to keep her warm, as the temperature was unreliable and at times tended to dehydrate her. So from then on she slept with us in our bed.

I returned to work on the Monday morning. No sooner had I reached the chimp house than the phone rang. It was the curator, Nick Ellerton, who asked me to come to his office straight away. Nick told me that Jane Goodall wanted to send me as a consultant to the Republic of Congo to help take care of some orphan chimpanzees in Brazzaville. Jane needed someone who had experience of hand-rearing chimps, could speak fluent French and was trained in anaesthetising animals. I fitted all these criteria: a year earlier I had attended a veterinary training course on anaesthetising zoo animals using a tranquillising rifle, pistol and blowpipe and I held a Home Office licence for prohibited weapons. Nick asked me if I was willing to go. This had all come out of the blue and I was unprepared, but I told him that I was interested, then asked when the trip was scheduled to take place.

'In three days' time.'

'What! You can't be serious. So when do I need to let you know by?'

Nick reflected for a few seconds before replying, 'I'll give you exactly one hour.'

I rushed back to the chimp house and phoned Audrey. I explained to her that this could be a useful career move for me and she agreed. She was also more than happy to be able to keep Sophie at home for a while. The next two days became a mad dash to get my visa and inoculations up to date.

2

The Front Line of Conservation

It was the first week of October when I boarded a Sabena aeroplane destined for Brazzaville. I had never travelled outside Europe before and wasn't quite sure what to expect or how this journey would pan out. The previous day Jane Goodall had briefed me on the purpose of my trip. She basically wanted me to assist in setting up the Brazzaville Chimpanzee Orphanage and to start the rehabilitation process with its orphan chimps. I was to be gone for three weeks, leaving a pregnant Audrey to look after Sophie on her own.

As I emerged from the coolness of the plane, the heat and humidity of Brazzaville struck me like a furnace. Jane had told me that I would be met at the airport but I didn't know by whom. So, as I walked through Customs, not knowing what to expect, it was a great relief to see my name written on a piece of cardboard held up by a tall, middle-aged Congolese man. He saw me look at his sign.

'Monsieur Vincent?' he enquired with a warm smile.

'Oui, c'est moi, bonjour,' I replied with a touch of relief.

'Aah, soyez le bienvenu au Congo.'

He was the chauffeur of the US Ambassador, Dan Phillips. He spoke only French, so this gave me the chance to practise mine. I was led to a smart Chevrolet car and then driven the short journey to Brazzaville Zoo.

* * *

The Republic of Congo is known colloquially as 'Congo Brazza'. This is to differentiate it from the neighbouring former Belgian colony of Zaire, which, until 1971, was called the Democratic Republic of Congo, or the DRC. The DRC was renamed Zaire by Mobutu when he became president in 1971. Mobutu's regime soon acquired a reputation as one of the most corrupt in African history. But later in 1997 Zaire reverted to its original name, following the overthrow of Mobutu by Laurent Kabila and his rebel army. A former French colony, the Republic of Congo is roughly two-thirds the size of France, and straddles the equator. The population, of just under three million, embraces several ethnic groups: the Bakongo, making up 48 per cent, the Sanga (20 per cent), the Bateke (17 per cent) and the Mbochi (12 per cent). One of the first things I discovered about this part of the world was the improbability of meeting anyone who could speak a single word of English. This was hardly surprising when you consider that most Congolese already speak at least two languages, French and Lingala, and usually another — their tribal language, such as Kikongo. So if you couldn't speak French, you'd be faced with a serious communication problem.

Five minutes later we arrived at the zoo, where I was warmly welcomed by the director, Mr Bukindi. I was shocked by the condition of the place. It was antiquated and the buildings dilapidated — essentially a ramshackle collection of concrete cages with rusty bars. During colonial times it had been used mainly as a transit holding place for animals being exported to Europe.

I was then introduced to Jean Mboto, who was employed by the Jane Goodall Institute and responsible for the welfare of the chimps. He gave me a tour of the zoo. There were 13 chimps (three females and ten males), most of them housed separately, a variety of monkeys, mainly baboons, mandrills, black mangabeys and a rare but very aggressive Wolf's guenon. Other than the primates, there were a pair of very undernourished lions, a red-river hog and a small herd of sitatunga antelopes.

The condition of the monkeys was a distressing sight. The bare concrete walls of their cramped cages were black with dirt and excrement. They spent their days staring out with outstretched hands from the rusty bars in the hope of receiving handouts from passing visitors — often their only source of food. They would while away the hours by catching and eating flies, which they managed to do with lightning speed as the insects swarmed around their faeces. One particularly skinny male mandrill was an expert. His hands were so fast I could barely see them move.

During my short stay in the Congo I experienced two extremes of living conditions. I was introduced to Mark and Helen Attwater and Steve Blake, who all worked for the John Aspinall Foundation, a charity run by the Howletts and Port Lympne Zoos in England. Both zoos were founded by the late John Aspinall, and between them they house the largest captive collection of western lowland gorillas in the world. Mark was the manager of the Gorilla Orphanage. The orphanages for both the gorillas and the chimpanzees stood within the grounds of the zoo. My hosts kindly offered to put me up at the gorilla orphanage for a few days. That evening we shared a couple of beers and swapped stories. Steve warned me that the previous occupant of the room, a young American student, had woken screaming in the night with a rat tugging on her hair. As I was soon to discover, Steve was renowned for his humour, and although I found his exaggerated story very amusing, I felt sure he was pulling my leg.

Soon everyone had gone home and I was the only human left in a building full of gorillas. My room turned out to be one of the spare gorilla beds in the orphanage. It was very basic — just four bare walls, a bed and a chest of drawers — but it was all I needed. However, there was one minor drawback: the room's only window was covered by a mosquito mesh which had a hole the size of my fist in it, and, as luck would have it, a pane of glass was broken. This small hole turned out to be the gateway for the inhabitants of the Congolese nocturnal world, and my first taste of Africa was soon to become Africa's first taste of me.

That night several uninvited guests decided to pay me a visit. Around midnight I began to hear loud scrambling and scratching noises coming from outside my window. These animals weren't insects or reptiles — they sounded too big. I lay motionless in my bed, trying not to breathe. The mosquito mesh began twanging, and this was followed swiftly by several loud thumps on the ground as whatever they were landed heavily on the floor of my room. Still holding my breath, I reached out and slowly took a torch from my side. As I gingerly shone its light, three large black rats came into view. Oh shit! I thought, as I jumped up and stood on my bed, sending them frantically scurrying in different directions. I have nothing against rats as a species so long as I know exactly where they are — preferably somewhere far away and not in my bedroom. I scanned the floor with the torch to see where they had gone, but they had vanished. It was then that I saw that the floor was crawling with cockroaches. I was alone in the building and had nowhere else to go. I felt the adrenalin rushing through my body and slipped back into my bed, pulling the sheets up tight around my face while I listened for more unwelcome sounds. My mind wandered back to Steve's tale of the screaming girl. Just how exaggerated this story was I wasn't sure, but as the night wore on it became disturbingly plausible.

Apart from worrying about losing my hair, I was also very conscious of trying to avoid malaria at all costs — especially cerebral malaria, which was rife in the region and could kill you in a few days. I shone the light on the walls and noticed a couple of friendly geckos. These at least were welcome and a little reassuring. They would eat some of the more unpleasant insects, I thought to myself. The numerous large cockroaches were not so well appreciated. Neither were the mosquitoes, with their distinctive hum, buzzing safely in the distance at first, then hurtling towards my face like crazed kamikaze pilots. My constant slapping of my face to ward off the advancing mozzies sent the many rats that were now in the room scurrying for cover. I tried sleeping with my head covered beneath my single sheet but the stifling heat and humidity made it impossible. In the end I was too exhausted from my travelling to really care. Outside my broken window, the resident bullfrogs began vociferously performing their evening concert. Their rhythmic, soporific tones drowned out the animal noises from inside my room, and I soon succumbed to sleep.

I was awoken early by shrill, ear-piercing screams from just behind my door. Feeling surprisingly refreshed after my punctuated sleep, I stretched out my arms and turned over slowly on to my side. As I surveyed the room I became aware that there was something small and dark very close to me. I moved my head backwards a little to focus my sleepy eyes. As I did so, a large rat dropping gradually came into focus. It had been positioned strategically on my pillow, inches away from my face. Flicking this calling card away, I leapt out of bed and quickly got dressed. The screams continued from outside my door. Slowly, I opened it, and peered into the corridor. A small, dark face with large, brown, sparkling eyes stared up at me inquisitively. It belonged to an orphaned baby bonobo, and he was impatiently demanding his breakfast. So I gave him a wave and then closed the door and proceeded to heat some water on a small camping stove.

Coffee tasted great and I was now eagerly looking forward to some breakfast. I had afforded myself the luxury of some food from the local Score Supermarket in Brazzaville. This was an expensive mistake, as I spent a fortune on virtually nothing, especially the Camembert. But I felt I deserved this little morsel of extravagance. I had been particularly careful to choose a ripe cheese, by gently pressing down on the wrapper with my thumb to test if it was soft. I had hidden the cheese the previous evening in the top drawer of the cupboard. This was the only place safe from insects, as the drawers fitted too tightly for anything to squeeze through. So I opened the drawer and was pleased to see that there were no creepy-crawlies inside. I will have to eat this soon, I thought to myself greedily, otherwise it will rapidly go off in this heat.

I opened the box and unwrapped the Camembert. I examined the cheese for a split second. Something was wrong; the cheese was black and moving. As a dark tide flowed up my fingers, I instinctively dropped the cheese to the floor, scattering millions of tiny black ants everywhere. Brushing the remaining dozens of them from my hands, I decided to pass on breakfast and went outside.

'Welcome to Africa,' I murmured softly to myself.

Jean was already in the kitchen preparing the food for the chimps. He looked up and smiled when he saw me. 'Bonjour, Monsieur Vincent. Vous avez bien dormi?' he asked politely.

'Oui, merci, Jean, très bien,' I replied diplomatically. He was so friendly, I didn't have the heart to tell him otherwise.

Unlike my Camembert, the chimps' breakfast looked delicious. It consisted of a variety of fresh tropical fruit such as papaya, mangoes, passion fruit, red tropical bananas and oranges, bread and a large bucket of milk. We set off with the food in a wheelbarrow and Jean introduced me to each of the chimps. I was impressed with Jean: he seemed genuinely to care for them. I got the impression that this wasn't just a job for him but that these were his friends too. As well as the ten youngsters, there were three adult male chimps: Zou Zou, Banane and Grégoire. Grégoire was a remarkable individual. He had arrived at the zoo as a baby and was now 49 years old. Although chimps can reach a half-century or more, this is exceptional. Not only had he survived to a ripe old age but he had done so under deplorable conditions.

When Jane Goodall first visited Brazzaville Zoo she was determined to do something to help Grégoire. At the time, he was terribly thin and almost completely bald. All his teeth were worn down virtually to their roots and he had the appearance of a withered old man. Now, thanks to Jane's efforts, he was well fed and his overall condition had drastically improved. Plans were under way to build Grégoire a large, outside run where he could live out the remaining years of his life.

That same afternoon a man entered the zoo and tried to sell a baby chimp to the authorities. They refused, of course, so the man instead donated the chimp. The owner told me that the chimp's name was Akim and assured me that he had been well looked after and hadn't been hunted. Akim was roughly six months old. He had a healthy, thick coat of hair and seemed well nourished. Apart from a small abscess on his lower jaw, his overall condition was good. I was then left on my own for at least an hour holding this wriggling baby who didn't want to be held by me, while the zoo staff tried to find a spare cage. Talk about being thrown in at the deep end. I'd been in Africa just a few hours and already I was left holding the baby. After what seemed like an eternity, they returned with a crate, and assured me that the many bite wounds across my chest and arms would heal quickly.

During my wrestling match with Akim, I had discovered that he had a shotgun pellet lodged in his chest, just beneath the skin. This was testimony to the circumstances of his capture: his mother had been killed as he clung to her. Using a scalpel and tweezers, I carefully made a small incision and removed the pellet, which I kept as a souvenir, a poignant symbol of the plight of wild chimpanzees and the way in which they are shot for food or commercial gain.

After a few days I met Lucy Phillips, the wife of the US Ambassador. She was visiting the zoo and bringing food for the animals. Lucy was appalled to hear that I was sleeping in one of the gorilla beds and kindly offered to put me up at her house. By this time I had repaired the hole in the mosquito mesh, so at least there were no more rats. In fact I was just beginning to find my room cosy. But I accepted her offer, and that evening she sent her chauffeur to pick me up from the zoo. The palatial ambassadorial residence, complete with gaudy marble pillars, seemed somewhat incongruous in Africa, but I must confess that I enjoyed Lucy and Dan's generous hospitality and the luxury of not having to worry about catching malaria. My bedroom, with en suite bathroom, was enormous. There was also air-conditioning, which made a refreshing change from my previous sweaty nights.

In fact the room was extremely cold — more like a butcher's coldroom. Before going to bed I had tried unsuccessfully to find the controls for the air-conditioning. It was freezing and impossible to sleep. In the end I could bear it no longer. I leapt out of bed and dug out a pair of socks and a jumper, which I pulled over my pyjamas. Still frozen, I curled up in the foetal position with my hands tucked between my thighs and my face buried beneath the blankets. Evidently, Lucy didn't like the heat. How ironic that the coldest night I could ever recall should be spent in equatorial Africa.

Breakfast the next morning soon had me forgetting about my night in the freezer, and I was now ready to get to work with the chimps. On opening the front door of the house I was immediately knocked back by the searing heat. Fortunately, Lucy had arranged for me to be chauffeur-driven to the zoo. The air-conditioning in the Chevrolet was slightly above freezing. When I opened the window it instantly felt like a hot hairdryer was being held to my face. I quickly wound up the window, electing for hypothermia rather than heatstroke. As we sped along, I pondered the wisdom of living in temperatures of extreme hot and cold.

Lucy was a very kind person who spent a great deal of her time actively involved with local women in the community. She had boundless energy and was always looking for ways to help others. Each day without fail she would come to the orphanage, ask how everyone was and then collect the chimps' dirty sleeping blankets ('dirty' being an understatement, as they were soaked in urine and faeces). Despite their condition, she would put them into the boot of her car and take them back to her house, where they were washed in her washing machine. They would then all be neatly ironed, ready to go back to the orphanage the following day.

The reason for ironing the blankets was not some quirky extravagance but necessity. One of the less delightful inhabitants of sub-Saharan Africa is a small creature known locally as the *tumbu* fly (*Cordylobia anthropophaga*). This insect has an interesting life cycle. The female likes to lay her eggs among the folds or creases on clothing, especially when it is lying outside to dry in the sun. The eggs hatch on contact with the skin and the emerging maggots burrow under your skin and stay there, gradually growing and feeding off your flesh until they are ready to emerge from the body of their host. At first you develop a small pimple but this gradually grows into a large, infected boil. The only sure way to remove these intruders from your body is to cover the infected sore with Vaseline. The theory is that the Vaseline prevents the maggot breathing, so it quickly burrows its way to the surface to find air. During my time here I had the dubious pleasure of removing several of these *tumbu*-fly maggots from many of the chimps. The locals always ironed their clothes — especially their undergarments — to ensure that all the eggs were destroyed. I was advised to make sure that my underpants were always well ironed.

One of the more memorable tales I was told was how a previous researcher, who was studying the gorillas, had been unfortunate enough to find a *tumbu*-fly maggot living beneath the skin of his scrotum. I winced at the thought, and after reassuring myself with a thorough inspection, enjoyed, for the first time ever, an evening's ironing.

The following morning I was greeted at the zoo by Graziella Cottman, a Belgian, who had recently fled the political conflict in Zaire, taking with her seven orphan chimpanzees. Graziella was in charge of running the chimpanzee orphanage but had been in Belgium for a few days visiting her daughter. She invited me to stay with her at her house. This was much more convenient, as she lived close to the zoo. So I thanked Lucy for her hospitality and moved into Graziella's more modest dwelling.

Together we set to work scrubbing and disinfecting the cages. There were 20 in all and we spent half a day on each. I arranged for metal rings to be made at the ironmonger's in town. These we bolted to the walls. Then we attached ropes and branches to them, and generally improved the chimps' living conditions. When the cages were all finished we sat back to admire our handiwork. It had made a big difference, and they looked less oppressive — at least we thought so, even if the occupants didn't. I suppose chimps aren't too bothered about aesthetics, but they must have appreciated the ropes and branches. Then we began introducing the infants to each other — one at a time. This was done in one of the cages. We spread out lots of food on the floor and sat in with them until we were satisfied that they had made friends.

Before the first week was over, two more babies, Pasy and Jacob, were donated to the zoo. Jacob seemed in good condition. However, on closer inspection I found he had a fungal infection and was covered in lice; it looked like a bad case of dandruff. Pasy was badly malnourished and dehydrated. He was also terribly traumatised, rocking backwards and forwards all the time, and helplessly lost if you took his blanket away, to the extent that he would collapse to the ground in a pathetic heap. It was sad to note that the owner of this chimp had been given a legal document by the Ministry of the Forest Economy, Fishing and the Environment entitling him to keep him as a pet. If that wasn't enough, Pasy also had a severe infestation of scabies. His chest, arms and legs were covered in these tiny parasites, which burrow beneath the surface of the skin. Despite having to hold the three chimps constantly in my arms, I managed to avoid infestation. Somehow I didn't think Audrey would see the funny side if I returned home and infected her with parasites.

We took a stool sample from each chimp and sent these away to the local laboratory. The results confirmed my suspicions: they were all infested with internal parasites as well as external ones. Akim had amoebas, Jacob ascaris worms and poor old Pasy had both. These young chimps were very demanding of my time, and it wasn't long before Akim in particular became clingy. During this desperate period in his short life he had chosen me as his adopted parent and wanted to be held throughout the day. I was used to carrying baby chimps because of Sophie, so it was second nature for me to do the same here. But there were times when I couldn't take him with me — in the evenings and when leaving the zoo at lunchtime. This meant creeping around the zoo, running from tree to tree so as not to upset him. If he did catch sight of me in the distance, he would scream frantically until he was hoarse.

By the end of the second week we had successfully introduced most of the chimps to each other and were able to walk them to a 50-acre (20 hectares) forest inside the zoo's boundary. Here they could climb and swing from the trees and thick lianas. They loved it here and it was satisfying to see them playing and having fun again despite all the traumas they had suffered. At first the chimps were nervous of leaving their cages and the three of us — Graziella, Jean and I — had to carry them. This sometimes meant having one on my back, another wrapped around my waist, two babies in each arm and one hanging from my leg. It was exhausting, especially in the stifling heat. After a few days the older infants grew in confidence and would walk in a long line to and from the forest.

One of the more boisterous males, Jay, who was about five years old, loved to climb high up into the trees above you and then dive on to your back when you least suspected it. Not only was this extremely annoying and painful, but I was constantly fearful that he might break my neck. Each day I would emerge from the forest battered, bruised and scratched and with my shirt buttonless and ripped to shreds. We gradually increased the chimps' time in the forest until they were only eating and sleeping in their cages. We were slowly preparing them for a life in a semi-wild sanctuary, which the oil company Conoco had agreed to fund and build for the Jane Goodall Institute near Pointe-Noire.

* * *

Away from the zoo, life in Brazzaville could be quite pleasant once you had adapted to the equatorial climate. Early on I had noticed how slowly the Congolese walked; they all strolled around at a snail's pace. It didn't take me long to fathom out why. The high heat and humidity sap all your strength. The city centre was relatively modern, consisting mainly of banks and offices. If you wanted to go shopping in Brazzaville, you would end up disappointed, as there was very little to grab the tourist's eye. In the evenings the locals and expats would relax in one of the numerous bars, while listening to the popular soukous music. Outside the centre were many half-constructed buildings, entombed in their wooden scaffolding, and along the roads lay the rusty shells of abandoned cars. The Congolese are a resourceful lot, so various parts of these cars would soon vanish overnight to be made into saucepans or other everyday utensils.

Brazzaville had a major problem disposing of its rubbish. The post-colonial infrastructure seemed to be crumbling, along with the pot-holed roads. Mountains of rotting waste were piled high on the very edge of the city centre, creating a health hazard. I had to walk past one of these dumps each day while I was staying at the Ambassador's residence. Many of the inhabitants lived in mud huts with rusting corrugated roofs, surrounded by tall palm and banana trees. The yards immediately outside the huts were always swept immaculately clean. The intense heat, the soukous music, the abundant palm trees and the deafening electric hum of mole crickets made the city feel very exotic, and all in all I loved the place.

Unlike its big-city cousin Kinshasa, Brazzaville has very little crime. The two cities are separated only by the Congo River, and Kinshasa, with its tall imposing skyscrapers, is clearly visible from the other side. One American journalist told me how he had left his wallet, containing several hundred dollars, in the back of a taxi. Incredibly, the driver drove around all day until he eventually found him and handed it back.

⋆ ⋆ ⋆

When Graziella went to the market each weekend to buy the chimps' food, I took the opportunity to accompany her. The main market was immense — as big as a small village — and you could easily get hopelessly lost. There were thousands of stalls and many thousands more people. The Congolese took great pride in their appearance: the men in their suits or smartly ironed jeans, spotless T-shirts and polished shoes; the women always elegant in their brightly coloured traditional dresses. The women would spend a day and a half at the hairdresser's having their hair braided; you had to admire their will power. In fact I felt conspicuous as the scruffiest person in the market, with my tatty, faded jeans, creased shirt and dusty shoes. The market itself was fascinating, as I was surrounded by various types of food that I was unfamiliar with. Even the smells were unusual, and often unpleasant — mainly because I didn't recognise them. At one point a man held up six dead fruit bats as I walked by. Their obnoxious stench — like damp, putrefying rats — left me feeling quite sick.

Much of the meat and fish was smoked, and so unrecognisable. A lot of the smoked meat was illegal bush meat. Mark Attwater had told me that they often saw gorilla and chimp body parts for sale here. The people believed that if you ate the meat of a gorilla, you would inherit its strength, and that by eating chimpanzee meat you would inherit the animal's wisdom.

Time passed quickly, and before I knew it three weeks had gone by. I recommended to Graziella that it would be useful to have a high-pressure hose for the project, as this was the only way to keep the cages truly clean. I promised that I would try to get one donated from England. Then it was time to say goodbye to the friends I had made at the zoo and head back to cooler climes. I had loved my experience here. From now on Africa held a nostalgic place in my heart, and I was determined to return one day with my family.

3

A Hairless Sibling

I returned home with many tales about Africa for Audrey, but before I launched into any of these she told me that Sophie had been very clingy while I was away. Sophie was upstairs sleeping peacefully in her bed when we gently woke her.

'Sophie, look who's home,' Audrey said softly.

Sophie slowly stirred from her sleep. She sat up on the bed and, with her eyes half open, looked across at me. I was expecting her to be overjoyed to see me but was surprised to see little reaction. Instead she just sat there looking at me, confused. At the time she was still very young, less than six weeks old, and I'd been away for half her life. Perhaps it was because she was still sleepy, but she didn't seem to recognise me. I felt a little hurt inside but tried not to reveal my disappointment to Audrey. I asked her how Sophie had been, but before she had time to answer, Sophie seemed suddenly to recognise my voice. She rushed across the bed and flung herself into my arms, gripping me tightly. It was a tremendous relief and a special and moving moment for me.

Our relationship continued where we had left off three weeks before. Audrey had been Sophie's nanny during my absence but hadn't replaced me as her 'parent'. So on my return her loyalties immediately switched back to me. From that point on we would become inseparable — something Sophie made sure of by never letting me out of her sight. I carried her wherever I went, just like a chimpanzee mother would do. I wanted her to become a mother and raise her babies herself. Many hand-reared apes do not rear their offspring themselves. I believe this is partly because they are often kept in small cages, where they are fed, cleaned, played with and occasionally carried.

A female chimpanzee's maternal skills are not innate — they need to be learned through observation and personal experience over a period of time. Most hand-reared infants do not experience adequate maternal care and consequently fail to learn crucial mothering skills. All too often babies of hand-reared mothers need themselves to be hand-reared, which perpetuates the problem. Primates generally take a long time to develop to maturity. An infant chimpanzee, for example, will often remain with its mother for up to five years. During this period it will develop many of the social skills necessary to allow it to progress smoothly into adult society. An adolescent female may get the opportunity to practise her maternal skills by carrying a younger sibling or the baby of a close ally. This is termed 'alloparenting' and is an important source of education for a young female chimpanzee. Hand-reared individuals usually don't get the benefit of this education. To my mind, one of the main reasons why so many hand-reared chimps don't raise their own babies is because when the infant is born they fail to pick the baby up from the floor. If they would only do this, then there is a fair chance that the baby would find the nipple and so bond with its mother.

As she grew rapidly larger and heavier Sophie became a burden on my left hip, which was suffering through my constantly holding her on one side. I tended to hold her in my left arm, and not only did she prefer to be held on my left side, she insisted on it. If my arm was getting tired and I tried to cradle her in my right arm, she would immediately position herself back on my left side. She quickly devised another way of getting a free lift, by holding on to my leg with her arms and parking her bum on my shoe. It was a bit cumbersome, like walking with a ball and chain around my ankle, and didn't do any favours to my sore hip. Again she would always choose my left leg. If I moved her over to my right leg she would immediately hop back over to my left foot.

Most human mothers tend to cradle their babies with their left arm. Humans, chimpanzees and gorillas share this feature. Orangutans and gibbons, however, tend not to favour any particular side. It is believed that left-side cradling is related to the different functions of the two halves of the brain. The human brain is composed of two halves, called cerebral hemispheres. The left hemisphere has specialisations for speech and language. The right hemisphere is important for functions such as interpreting the emotional significance of both facial expressions and sounds. Most of the brain's visual and auditory fibres cross over from the left eye and left ear to the right hemisphere of the brain, and vice versa. It is believed that a mother can better judge the emotional state of her infant by positioning herself more on its left side. There is evidence to suggest that an infant also gains advantages by being cradled with the left arm, as it then has a better view of the emotionally expressive side of its mother's face.

Interestingly, in a study of 300 humans who were asked to imagine holding certain objects, including a baby, right-handed adults tended to cradle the infant in their left arm and hold objects in their right hand — their stronger hand. Left-handed adults tended to cradle the baby in their left arm but also use their left hand — their stronger hand — for carrying objects. Such 'brain lateralisation' appears to exist in chimps and gorillas — as mentioned, most cradle their infants on their left side, whereas orang-utans tend to show no preference. This provides more evidence to support the theory that humans and the African apes (gorillas and chimps) branched off from the human evolutionary tree later than orang-utans, and further demonstrates the evolutionary closeness of humans and the African apes.

I had to find another way of carrying Sophie, otherwise I was going to end up permanently walking like a chimp. So I hit upon the idea of using a baby sling. This worked a treat despite causing a few raised eyebrows among my colleagues. It was less awkward and a tremendous relief to my back, as she was now positioned centrally to my waist. I was at last able to walk like a human with my back straight instead of fixed at an angle to one side.

Back at home, Audrey was developing a different kind of back pain. Her pregnancy was starting to show. Being pregnant suited her and she was now proudly wearing smocks around the house. She was very happy and looked positively radiant. We would lie in bed in the evening and take turns to feel our baby moving around inside its aquatic world. Audrey would take my hand and place it on her stomach. Our baby had a firm kick.

'It must be a boy,' she would say, 'and I think he'll be a footballer.'

We would while away the evening hours thinking up names for our baby.

Audrey began attending pre-natal classes, and I would drive her to the hospital and then wait for her in the car park with Sophie. I managed to attend two of these classes by arranging for a neighbour's daughter to babysit Sophie at home. Audrey also went for regular check-ups at the maternity hospital. When the foetus was about six months old, I accompanied her on one of these visits. It was an anxious time for both of us as we waited at the maternity ward for encouraging news on the progress of our unborn child. And it was always a relief to be reassured that our baby was doing well. The nurse allowed me to join Audrey as they performed a scan of her tummy. The foetus was already a perfectly formed baby, although the nurse informed us that there were as yet no fingernails. These would develop later, she reassured us.

'There's the baby's heart beating,' she added.

I was sure that I saw our baby give me a wink and a thumbs up, although maybe he or she was just lifting a hand to suck the thumb. Although the nurse probably knew the answer, we chose not to discover the sex of our baby, being perfectly contented with either a boy or a girl. Although there are practical reasons why one might want to know this information — for buying appropriate clothes, for example — we didn't want to spoil our surprise. I wanted a little girl to cuddle but at the same time I wanted a boy, so that I could teach him how to play football and have an excuse to play with his toys. We returned home proud and happy with a photo of our unborn baby to show Sophie. She wasn't too interested in our photo, but was pleased to see us back. That evening we decided that if the baby turned out to be a girl we would name her Sabine, and if it was a boy we would call him Oliver.

Meanwhile Sophie was continuing to grow at a rapid pace and had now outgrown my home-made nappies. She was at long last big enough to wear the smallest size of nappies available in the shops. Hurrah for Pampers — and what a wonderful luxury! No more fumbling about in the middle of the night with a bag full of poo.

Because we couldn't leave Sophie alone in the house, we had to take her with us wherever we went. When she was five months old, Audrey and I went shopping in Shrewsbury. Normally one of us would stay at home with Sophie, but Audrey was now close to giving birth and couldn't carry heavy shopping alone. Sophie was asleep in the back of the car, wrapped in blankets, with just her head peeping out. While we were waiting at the traffic lights in the high street, one of Audrey's friends happened to be walking past. She hadn't seen Audrey for a long time and didn't know about Sophie. On spotting us, she ran over to find out if Audrey had given birth yet. She looked into the car and saw there was a baby in the back. Her excited smile was rapidly replaced by a look of horror. As Audrey was trying to wind down her window to explain that she hadn't given birth to a hairy baby, the lights turned green and I was forced to drive off. Anxious that everyone would be talking about her, Audrey had to telephone all her friends the following day and tell them about this other baby in our lives.

On the morning of 8 April 1991 Audrey's waters broke and she began having contractions. Sophie was now six months old and I had to leave her alone in the house for the first time so that I could take Audrey to hospital. Audrey calmly packed her bags and waited for me in the car. I took Sophie into our bedroom and played with her on the bed for a short while. I brought lots of toys and some fruit with me. Then, when she wasn't looking, I sneaked out of the room and locked the door. As soon as she heard the door close she cried out and rushed over to it, frantically turning the handle back and forth. I ran downstairs and jumped into the car. As we drove off, we caught sight of Sophie's face pressed against the upstairs window, as she screamed and pleaded for us to come back. We both felt terrible, but there was little else we could do as there had been no time to arrange for a babysitter, and I don't think Shrewsbury Hospital would have taken very kindly to her hairy presence in the labour ward. Audrey was about to give birth and yet she was more concerned for Sophie than for herself. She told the doctors and nurses about her and asked if it would be all right if I brought her into the ward. Understandably, they said no, but wrote on her medical notes about her concern for a chimp, or was it their concern for her sanity? Anyway, I doubt if many pregnant women have had similar information added to their medical records.

I remained with Audrey for the several hours that she was in labour. Throughout this time she constantly squeezed my hand and breathed in gas and then, late that afternoon, she gave birth to a healthy baby boy. He weighed six pounds nine ounces (three kilograms) and already had an impressive head of dark hair. In fact Audrey, in her drowsy state, thought she had given birth to a baby chimp. The two of them remained at the hospital and I returned home a proud father. I went upstairs and found Sophie fast asleep on the floor next to the bed and could see she hadn't played with any of the toys or touched her food.

On his second day Oliver developed mild

jaundice and had to be kept in a special incubator. It was very upsetting for us to see our newborn baby unwell and he looked pitiful lying there on his own, with his name tag hanging loosely around his wrist. Naturally, Audrey wanted to hold him in her arms. But we were reassured that it was quite common for babies to have this problem and the next morning he was back in bed with his mother.

A few days later I had to leave Sophie alone in the bedroom again, as it was now time to bring my family home to Meadowtown. Once more I had to trick her in order to lock her in the room. This time she sounded even more upset, realising from her previous experience that I was likely to be gone for a long time.

When we returned home an hour and a half later, Sophie was asleep in exactly the same position as I had found her before. We woke her up and Audrey gave her a big cuddle. Sophie was pleased to see Audrey and gave a little hoot of excitement. Then we carefully introduced Sophie to little Oliver, not knowing whether she would be jealous. On the contrary, she was very gentle and fascinated by the new member of the family. Now we were four. I believe she wasn't jealous because she didn't see Oliver as a threat to her. After all, I was Sophie's mother and Audrey was Oliver's.

Thanks to Sophie, we were already well rehearsed in looking after a baby. Audrey would change Oliver's nappy most of the time, and I continued to change Sophie's. Oliver was breastfed by Audrey and was a contented baby who hardly ever cried. In fact both babies seemed very cheerful, as they received food whenever they wanted and had our constant attention. When Oliver started on solid foods, we would take it in turn to feed him. Curiously, when spoon-feeding either Oliver or Sophie, I was always unable to suppress a kind of 'feeding reflex', opening my own mouth wide in typical feeding fashion at the same time as the babies.

Seeing Oliver and Sophie lying side by side was interesting. Oliver had as much hair on his head as Sophie, but her back, arms and legs were covered in thick, black hair. Although primates are covered in dense hair, the chimpanzee actually has fewer hair follicles than we so-called hairless humans. We 'naked apes' are covered in tiny fine hairs. We wrongly view a primate's body hair as a primitive trait. One slightly unconventional theory — the aquatic ape theory — speculates that after our primitive ancestors and the chimpanzees separated from a common lineage, they lived a semi-aquatic existence beside estuaries, marshes and the sea.

This intriguing theory has much evidence to support it. There are several characteristics humans share with aquatic mammals. Hairlessness is fairly uncommon in most land mammals and is unknown in other species of primate, but is, however, found in many species of aquatic mammal, including dolphins, whales, seals, manatees and hippos. Hair or fur is important for maintaining body temperature, and so is subcutaneous fat, which helps to insulate aquatic mammals. Humans are the only primates to have fat over their entire body. This layer of fat is attached to the skin, as it is in other aquatic mammals. However, in most land mammals it is attached to the muscle. Like aquatic mammals such as the walrus, sea lion and manatee, we have a descended larynx. This gives us the ability to control our breathing, as it allows air to be taken in through the mouth, in deep breaths, and held while underwater. At birth human babies are largely helpless. Even so, they are born with the ability to swim. They can also float and hold their breath underwater. Unlike when it is on land, the human baby is not helpless when in water and can swim beside its mother.

Innate traits are often a reflection of a previous lifestyle during an animal's evolutionary history, so this ability to swim at birth may indicate that at some period in our evolution we lived a more aquatic existence. One has to wonder if we would perceive our ape cousins a little differently if our body hair were thicker and longer.

Sophie and Oliver were generally both healthy babies who were rarely ill. Often their only ailments — slight colds or mild diarrhoea — were during teething. Their gums would become swollen and sore as new milk teeth erupted. We would treat them both with a spoonful of Calpol — a paracetamol for children in the form of a syrup. Sophie's milk teeth began erupting before Oliver was born, when she was less than three months old. These were her front central incisors, followed later by her lateral incisors. It was very gratifying to see Sophie with her gleaming white teeth, and each day we would check on their development. We would ask her to open her mouth and she would gladly oblige. I suppose to her it was a form of grooming and she didn't seem to mind our regular inspections.

Soon after Oliver was born, Sophie developed a lactose intolerance to her powdered cow's milk — something that is quite common in children. This led to her having diarrhoea which persisted for quite a while with no sign of improvement. Eventually we were forced to change her regular milk diet to a soya-based powdered-milk formula. This seemed to do the trick, as the diarrhoea stopped soon afterwards. We kept her on a soya milk diet for about a month and then gradually reintroduced her original SMA lactose milk. We did this by initially feeding her milk heavily diluted with water and then gradually increasing the proportion of milk.

As the two babies grew up together, they became inseparable friends. They had the same parents, lived under the same roof and shared similar experiences. But they were much more than just friends — they were part of a family.

4

Actions Speak Louder than Words

Like Sophie, Oliver had a powerful grip for the first few days of his life. I remember placing my index finger into the palm of his hand as he lay on our bed and being surprised by his strength. As I pulled my hand away, he clung on firmly so that his body was lifted up to a sitting position. All healthy human babies are born with a powerful grip; during the first few days after birth, a baby is able to grasp objects with its fingers and toes. So strong is this grip that the baby can support its own body weight with one hand, although this ability gradually diminishes as it gets older. This powerful grip possessed by our hands and feet is believed to be a primitive trait retained from a period during our evolutionary past, perhaps when our primitive ancestral infants needed to grasp their mother's hair for support as she moved through the trees.

During the first few months Sophie was much more advanced physically and mentally than an equivalent-aged human child. In chimpanzees, the rate of brain growth slows down shortly after birth, but continues in humans throughout their first year, so it was only a matter of time before Oliver would catch up and overtake Sophie. Forever curious about her younger companion, she would constantly check what he was up to. She seemed to be willing him to grow up and join her on her explorations. One day when we were downstairs in the kitchen, we heard a loud bump from upstairs, followed a few seconds later by the sound of Oliver crying. I rushed upstairs to find that he had fallen off the very same bed that Sophie had tumbled off a few months earlier. Oliver had learned to crawl and, like Sophie before him, could now no longer be left asleep alone in bed. Now that both of them could crawl they were much more compatible as playmates.

Sophie was a tremendous influence on Oliver's development. Oliver would try to copy much of what she did. We bought him a playpen in which he would sit quite contentedly, and out of harm's way, with all his toys. Sophie would constantly climb in and out to play with these, and it wasn't long before Oliver decided that anything Sophie could do he could do too. Soon he was copying his big sister. At first he would climb up the bars and then struggle to get his leg over the top. But practice makes perfect and soon he had the hang of it and was climbing out at will. The next annoying thing that Sophie decided to teach Oliver was navigating the stair guard. Much to our concern, Oliver was soon climbing over this with ease and scrambling up the stairs after Sophie.

The two of them developed their own little games. One of their favourites was a form of tag invented by Sophie. Oliver would lie on the bed and dangle a hand off the edge, and Sophie, who was hiding beneath, would try to grab it. Then Sophie would push her hand up through the gap between the wall and the bed and Oliver would try to do the same. We soon knew, from their spontaneous laughter, if either had succeeded in tagging the other. Although incapable of human speech, chimpanzees love to laugh, especially when young. Sophie enjoyed being tickled, especially on her chest and neck. She would begin with a soft, rasping 'hi-hi-hi' sound (pronounced like a hissing snake) and end with a loud, raucous belly laugh that was so infectious you couldn't help laughing yourself.

I could see that Oliver was going to be a real comedian when he grew up. Even though he hadn't learned to say his first word, he already had a sense of humour. One memorable moment took place in the kitchen. Oliver was dancing on his own to music on the radio, when he noticed me laughing at him. He found this equally amusing, and then kept repeating the same action to see if I would laugh again, which I found even funnier. On another occasion he started playing peek-a-boo in order to make me laugh. I had always believed that it was the adult who was supposed to make the baby smile, and not the other way round.

Thanks to Sophie's expert tutelage, Oliver became the first child to stand up in his mother-and-baby class. This wasn't surprising, as he had already learned to climb before he could walk. Fortunately, Sophie didn't get as far as teaching him the joys of swinging from our lounge curtains. It was interesting that Oliver was imitating Sophie and not the other way around. Contrary to conventional belief, it is humans, and not apes, who have the greater ability to 'ape' another's behaviour. In a study on how children and chimpanzees communicate and learn from others, it was found that human children were much better at imitating than chimps. Children tended to copy the way that adults performed certain actions, even when the techniques used were relatively inefficient. The chimps, by contrast, tended to observe the action and then tried to improvise their own strategy. Although the chimps' approach appeared more creative, the children were taking advantage of important skills and knowledge learned from others.

Sophie had become as much a part of our lives as any adopted human child. She had our constant attention, day and night. At times this created a problem, as she wasn't independent enough to be left alone. It wasn't surprising, though, as I knew that a chimpanzee in the wild normally stays with its mother for up to five years. During that important period of development the mother is the main focus of the infant's life.

Most of the time, if Sophie wasn't clinging to my waist, she would be playing close by. She would always stay within sight of me, even when we were in the house. But I found it interesting that when we visited friends in a house that she had never been to before, she would become much more confident and happily disappear upstairs to explore. Then she would come back down every now and again to check on my whereabouts. This was very much the case if there were young children present. It was as if she knew we would never leave her in a foreign place, but might well do so in our own home — as we had done previously. We always tried to treat her as we did Oliver. After all, we rarely used a babysitter for Oliver, and wanted, as much as possible, to take Sophie everywhere with us too.

* * *

One day I decided to take the family out to Shrewsbury for a meal. I hadn't left Sophie alone in the house since fetching Oliver from the hospital. So I took Sophie upstairs to lock her in the bedroom. However, she could sense from the activity downstairs that we were planning to go out. When I took her upstairs she suspected I was up to something. I put her down on the bed, but before I had time to close the door she had rushed past me, whimpering. She ran downstairs, through the lounge, the dining room, the kitchen and out of the back door. We found her sitting in the back of the car waiting for us — still very upset. She had opened and closed the car door behind her. We couldn't believe it. Faced with such a powerful statement, what else could we do but take her with us? So we walked into Pizza Hut in Shrewsbury with Sophie tucked under my jumper. A waitress came over and, after asking if we wanted a smoking or non-smoking table, led us to our seats. Every now and again this large lump would jerk about beneath my jumper. But the waitress didn't seem to notice, or at least she didn't let on if she did.

We sat down and I somehow managed to keep a rather agitated chimp still while we ordered our food. After a short while Sophie smelt the pizza and there was no holding her back. My chest was transformed into that infamous gory scene from *Alien*. Eventually the smell proved too much of a temptation and there was nothing anyone could do to prevent Sophie from erupting from my jumper. She emerged hot, bothered and very hungry. So I sat her in the corner between the wall and me, where she remained on best behaviour. None of the staff ever realised that sitting among us, tucking into a slice of Pizza Quattro Stagioni, was a small, hairy chimp, looking just like a gremlin. Oliver, of course, thought the whole affair was totally normal. After all, he had spent his entire early life with a chimpanzee.

Despite growing up with a chimp, Oliver was never particularly interested in our other animals. Sophie, however, was especially fond of Sabby but disliked her two sons, Tinge and Tufty, two kittens we had kept from her first litter of six. In her unsubtle way she would rush up and grab Sabby and cuddle her. Sabby would visibly grimace yet reluctantly put up with this unwelcome affection. Sophie wasn't quite so forthcoming with the young males and wouldn't tolerate them near her. In fact she would chase them away — maybe because they never fully accepted her and were nervous in her presence.

Chimps are natural bullies, and despite her being so young, Sophie found the kittens' fear of her a tremendous boost to her chimp ego. So chasing them became as much fun as the idea of cuddling them.

One thing Sophie tolerated less than the cats was anyone touching me — something that immediately provoked her jealousy. Other than herself, no one but Audrey and Oliver was allowed the privilege of touching my sacred body. Among primates, the infant's relationship with its mother is probably the most important one it will ever have. The mother provides security, nourishment, education and comfort for the infant. Sophie was determined to protect her own interests and was not going to share me with anyone or let anyone harm me.

She always wanted to learn as much as she could about her surrounding world. From an early age she closely observed everything I did and often attempted to imitate me, whether I was opening a can, taking the cap off a bottle of beer, turning on the tap or simply switching the lights on or off. I remember some friends remarking on how intelligent Oliver was because he was trying to use a screwdriver to undo screws on our fridge — and yes, Oliver probably *was* advanced for his age. Yet Sophie had started undoing screws with a screwdriver several months earlier. If I was busy writing letters or reports, she would come and sit quietly by me and study intently what I was doing. Then, after I had finished, she would pick up the pen and start drawing on a piece of paper. Interestingly,

Oliver's early attempts at drawing were almost indistinguishable from Sophie's; both consisted of disjointed lines and scribbles. One of Sophie's works of art adorns the back cover of the 1991 chimp diary at Chester Zoo. Her new-found penchant for drawing later became a minor nuisance, especially when I came across her scribbling in one of my bird books.

Another of Sophie's observations concerned the whereabouts of our food store. As a baby, she was never a very accomplished thief. She would succeed in sneaking into the kitchen unnoticed, open the door of the fridge, carefully remove some choice item and creep away. But, at the last minute, she would give herself away in her excitement, as she let out soft 'ah, ah, ah' sounds. She could never suppress these food grunts, and if we were in another room and heard them we immediately knew that she was up to no good. Oliver was quick to realise that anything Sophie could do, he could do better. Raiding the fridge *without* grunting was one of the more rewarding things he learned from her.

Oliver was growing up fast, and among his first words were 'Mamma', 'Dada', 'no' and 'cheese' — he was especially fond of cheese! As a rule, Sophie and Oliver were remarkably well behaved and never any trouble at all. But, as with any child, there were times when I had to put my foot down and say no. Sophie was always very responsive and usually stopped if I told her off. But I soon discovered that she responded much quicker to the chimpanzee version of the word 'no', a guttural 'uhh!', delivered in the form of a loud, short cough. Sophie was always very careful when jumping on to the table — something we didn't encourage — and seemed instinctively to know where to place her feet even when in a hurry. At times it was as if she had eyes in the back of her head. She rarely broke a glass or plate. On the rare occasions that she did — for example, when clambering over our collection of crystal glasses — it was normally because we had shouted, 'Sophie, uhh!' This would result in her hurriedly turning around, wondering what was the matter and losing concentration. Her amazing acrobatics could be accounted for by the fact that chimpanzees, having a tree-dwelling lifestyle, need to be more accurate when judging distances than us more cumbersome humans. 'Uhh' was so much more natural a sound to produce than the human 'no' that in times of crisis I would unwittingly react by using it with Oliver, the cats or even our chickens, much to the amusement of Audrey and friends.

Although chimpanzees are incapable of complex speech, they are nevertheless able to communicate through a simple repertoire of sounds, such as pant hoots, lip smacks and various grunts, and through non-verbal gestures. They vary the pitch, rate and volume of their vocalisations to convey different meanings in a variety of circumstances. The chimpanzee ear is similar to the human ear and can accurately differentiate far more sounds than the animal can produce. Unlike humans, chimpanzees lack the necessary anatomy for complex speech. In the chimpanzee, as in most mammals, the larynx is high up in the throat and serves as a valve to stop water entering the windpipe when drinking. It is believed that some 100,000 to 150,000 years ago the human larynx descended, creating a gap in the throat, termed the pharynx or voicebox. This allowed sounds to be resonated, which in turn increased the clarity and variety of sounds that could be produced. The tongue also altered, becoming thicker and more muscular, and this allowed more rapid and precise control of the tongue's shape, which further broadened the range of sounds produced.

Today humans and chimpanzees are born with a larynx that is set high in the throat, which allows them to drink and breathe simultaneously. But humans lose this ability by the time they are 18 months old, because the larynx descends, creating a common food and air passage, whereas the chimp larynx remains permanently in its original position. The position of the human larynx can cause a person to choke to death, whereas this doesn't happen in animals that can breathe and swallow at the same time.

Although structurally similar, the human brain is about twice the size of that of the chimpanzee. A large brain is fundamental for the production of complex language. The human brain is asymmetrical. The left side is slightly larger than the right and is believed to control language. There are two areas on the left side of the brain, Broca's and Wernicke's areas, which are known to be responsible for producing and comprehending language respectively. Chimpanzees appear to lack Broca's area in their brain, but have Wernicke's area. This may indicate that, although chimpanzees are unable to produce vocal language, they appear to have the capacity for processing complex language — maybe not verbal language but a non-verbal one consisting of sounds and body gestures.

The chimpanzee's vocal repertoire is limited compared with that of humans. However, much of their communication is achieved through gestural signals. Humans also use gestures in their everyday speech. These allow extra information to be passed — for example, in the classic fisherman's tale of the fish that got away. Language may have emerged not from vocalisations but from manual gestures, and switched to a vocal mode relatively recently in hominid evolution, perhaps with the emergence of *Homo sapiens*. It is interesting that human babies are able to make complex gestures before they have learned to speak, and children who make referential gestures at an early age tend to speak at an early age.

Attempts have been made to teach apes spoken language, but most have failed. However, despite not having the capacity for complex speech, apes are capable of learning complicated human sign language. Several apes have been trained to use the American Sign Language (ASL) and can communicate in this way with other apes or with humans, using several hundred signs. The fact that apes have shown the capacity for learning our complex human sign language indicates that they have the intelligence to learn our spoken language. It also underlines the value of focusing on their strengths rather than their weaknesses, for although apes may lack the anatomy for complex speech, they don't necessarily lack the intelligence.

Washoe, a female chimpanzee, was raised as if she were a deaf human child, and taught ASL. By the time she was four years old she had acquired a repertoire of some 130 signs. Washoe displayed the ability to think abstractly, make her own combinations of taught words to form new ones and apply simple rules of grammar. She adopted a ten-month-old male chimp called Loulis and taught him several signs, such as 'sit', 'food' and 'come'. Later Washoe and Loulis were joined by three other ASL-signing chimps, Moja, Dar and Tatu. In one study these five chimps were observed to sign to each other on 5,200 occasions.

Apes other than chimps have also been successfully taught human sign language. Koko and Michael, both lowland gorillas, developed an impressive vocabulary of ASL signings, Koko mastering 500 different signings and Michael (who, sadly, died recently) 400. Chantek, a male orang-utan, has learned 150 ASL signings.

These apes have all demonstrated the ability to use signs for intentional communication in a variety of ways. They have been observed to teach other apes to sign. When signing they take the initiative rather than just responding to questions from humans. Their signings tend to be spontaneous and non-repetitive, and are often referential, naming something in their environment — for example, a dog or car — just for the sake of it. They invent their own signings and combine two or more signs to form novel ones in order to convey different meanings. Among the many invented signs are the following examples: Washoe signed 'rock berry' for Brazil nut and 'water bird' for swan; Chantek signed 'eye-drink' for contact-lens solution and named a human 'Dave-missing-finger' for obvious reasons; Koko signed 'bottle match' for a cigarette lighter; and Michael signed 'bean ball' for peas. Of 876 signs made by Koko, 54 (6 per cent) were her own inventions. A further 2 per cent were her own combinations of taught signs.

Apes have also been taught to communicate through computerised keyboards. In 1979 Sue Savage-Rumbaugh, Professor of Biology and Psychology at Georgia State University and one of the world's leading ape-language researchers, began training a bonobo named Matata to communicate using symbols on a keyboard. But Matata wasn't the best of students, and learned very few of the symbols, possibly because she was adult when she started language training.

However, her adopted son, Kanzi, seemed fascinated by the keyboard, and had been closely observing the proceedings. He astonished his trainers by correctly using the keyboard to name items and to announce his intention. After 18 months Kanzi had learned a vocabulary of some 50 symbols and was regularly using combinations of words. As an adult, he has mastered over 200 words and is able to form sentences of three symbols strung together, sometimes in novel combinations, demonstrating a grammatical capacity that approximates to that of a human two-year-old.

Although these apes have shown the capacity for learning sign language or for using symbols to communicate at a level comparable with a two- or three-year-old human child, they are generally incapable of going beyond this basic level. Children, by contrast, go on to acquire a sophisticated language that is distinct from all other forms of animal communication in being generative — capable of being produced in an infinite number of variations — and highly flexible. Much of this is achieved through the use of syntax and grammatical structure, allowing us to create an infinite number of sentences of any desired complexity.

Although Sophie was unable to communicate by way of complex speech, she was very adept when it came to non-verbal communication, and was an expert at reading body language. Non-verbal signals are normally used to convey emotions and moods. One time when I was leaving the zoo to go home, I realised that I had left a book in the building. So I left Sophie in the car and popped back into the orang-utan house, foolishly leaving my car keys in the car. Sophie was fine about my leaving her, as she knew that I would return. When I came back a few moments later and was approaching the car, I saw that Sophie's foot was very close to the door lock. Before I could get to the car, she unwittingly stepped on the knob and centrally locked all the doors. She was now locked inside and I was locked out. Sophie remained perfectly happy until the moment she saw my worried expression. Intuitively she knew that something was terribly wrong, and began screaming hysterically. I tried to calm her down, but she wouldn't stop.

Leaving her screeching, I ran back into the orang-utan house and managed to find an old coat hanger. With Sophie still frantic, I opened the door by slipping the wire through a tiny gap in the window and hooking it around the top of the knob. Phew! Sophie had just been given a clean nappy for the journey home and now it was very, very full. As she grew older, getting locked in the car ceased to be a problem, as she simply unlocked the door herself.

On another occasion I was feeding the Chester chimps with Sophie in my arms. One of the adult males, Friday, was looking intently at my shoe, but in a menacing way. I could see that he was waiting for an opportunity to grab hold of my foot. I was surprised to see that Sophie had also accurately read the situation. In a flash, she wriggled free from my grasp and leapt down to the ground, barking and waving her arms angrily at Friday through the bars. I had to take her away for her own safety, as Friday would have readily grabbed her instead of my shoe.

As Sophie grew older, so did our popularity, and we had a constant stream of visitors to the house. Sophie loved greeting and meeting visitors and would make a big fuss over them. She was the centre of attention, the life and soul of the party, and she knew it. If she heard a car approaching the house she would announce its arrival with loud hoots of joy. In the wild, chimpanzees communicate with each other over great distances by emitting loud pant hoots. These sounds convey a variety of information, and they use them to tell group members of their whereabouts or the presence of fruiting trees, or to warn them of invading groups of chimps. But they can also mean, 'Friends are coming — get the tea and biscuits'!

'What a cute monkey,' visitors would say, much to Audrey's annoyance. How many times she had heard people say this she couldn't remember. It was like an insult to her child, and she would feel compelled to correct them. 'Actually, she's not a monkey, she's an ape.'

'Oh really, what's the difference?' they would ask politely.

This was a very pertinent question. After all, the differences between apes and monkeys are not that obvious to the untrained eye. However, apart from tending to be larger in body size than monkeys, apes have the following distinguishing physical characteristics: distinctive molar teeth, with a five-pointed crown; a semi-erect posture; highly flexible arm and shoulder joints; arms longer than their legs; and a larger and more complex brain than that of a monkey. More noticeably, they have no tail.

During development, the human foetus has what is termed a vestigial tail. This is visible by the fourth week of development and reaches its maximum length at the sixth week. It then gradually shortens until, by the time the baby is born, it becomes a vestigial organ called the coccyx. Very rarely a human is born with a structure that resembles a tail. Vestigial organs tend to serve no purpose, but as they resemble functioning organs in other animals, they suggest a common ancestry. Evolutionists, much to the annoyance of creationists (who promote the theory of divine intervention during human prehistory), often use vestigial organs in developing human embryos as evidence of a common ancestry between humans and primates.

Visitors were always welcomed by Sophie. She loved being picked up and cuddled by our friends, especially by children, with whom she had a particularly strong affinity. She reacted differently to the various age groups, being rough and boisterous in her play with adults, less so with children or the elderly and extremely gentle with babies. I was curious to see how she would react to meeting members of my own immediate family, my mother, Paulette, and my two sisters, Madeleine and Christine.

Well, I didn't have to wait long to find out. To celebrate Christmas in 1991 my mother had arranged a family get-together at her house in Shepton Mallet, Somerset. Audrey's mother, Solange, was also with us. Sophie got on well with Solange and was always very gentle with her. Both Solange and my mother had first met Sophie when she was three months old and still living in her cardboard box. However, unlike with Solange, as Sophie grew older she refused to allow my mother to pick her up, no matter what efforts she made in this direction. Sophie wasn't fearful in any way, just uninterested and aloof. If I picked her up and then tried to pass her to my mother, she would squirm out of her arms and scamper off.

It was Christmas Day and this was the first time that Sophie had met my sister Madeleine and her husband, Manuel, who lived in Spain. She got on fine with their son, Gael, but for some reason didn't want much to do with Madeleine or Manuel. Instead she had great fun tormenting them, to such an extent that the day turned into a disaster. Sophie, however, had a wonderful time, made all the more enjoyable by the teasing she forced my family to endure. Throughout our Christmas dinner she insisted on sneaking under the table as we ate our turkey and biting Madeleine or Manuel firmly on the toes or ankles. Not enough to draw blood, which she was quite capable of doing, but enough to get them leaping out of their chairs. By the time the meal was over they were both very stressed out. Despite being so small, Sophie knew that with me around she could get away with almost anything. All this teasing soon tired her out, though, and later that afternoon I found her lying on Solange's lap in the conservatory. Both of them were sound asleep. This was the first and only time I've ever seen Sophie sleep on anyone other than Audrey or myself. Sophie and Solange were both born on the same day, so maybe their close affinity had been preordained. Fortunately, with Sophie asleep and out of harm's way, peace gradually returned to the household, and we were able to enjoy the rest of our evening.

Infant chimpanzees have a tuft of white hairs on their rear. These hairs are present until the infant is four or five years old, and act as their carte blanche to do whatever they want. Adults are very tolerant of infants with these white hairs, and the youngsters know it. Basically, they are allowed to get away with murder. For example, knowing that their mother is close at hand, they will often pick an argument with others much larger than themselves. However, there comes a time when these hairs have all but disappeared, and if they step out of line then, their transition into adulthood can be a rude awakening. Sophie still had her small white tuft, although it was not as prominent as Kaylie's, possibly because her nappy had rubbed some away.

When Sophie finally met my other sister, Christine, there was an even worse reaction. Christine was living in France. She loved animals, probably even more than I do, and was excited about meeting Sophie for the first time. Sophie was playing alone in the dining room when Christine approached, knelt down in front of her, stretched out her hand in friendship and said, 'Hello, Sophie.' After stopping what she was doing, Sophie calmly grabbed Christine's hand and bit it hard. She wouldn't let go and I had to prise her mouth from my sister's hand. Although, once again, she didn't draw blood, I could see by the deep, blue teeth marks that it must have been painful. Sophie quickly gained considerable and immediate respect from my family. But I was surprised by her negative attitude towards them. She never reacted like this to anyone else — only to my family or to someone she felt was trying to harm either her or her family. This was intriguing, although rather embarrassing for me, but, of course, it was disappointing for my family. My personal feeling is that Sophie was observing the body language between my family and me and, being unfamiliar with this, was maybe interpreting certain postures as being threatening.

Although Sophie was generally a very gentle individual, anyone who crossed her had to beware, for she had a feisty temper. One of my colleagues at the zoo discovered this to his cost. I was in the chimp kitchen with the rest of the team, Neil Ormerod, Steve Hogarth, Ross Meredith and Steve Cook, when Sophie somehow managed to get herself tangled in the phone cable. It had become wrapped around her neck twice, and her attempts to free herself only tightened it more. Steve Cook saw her struggling and went over to assist her. But he made the mistake of going up behind her to try to remove the cable. Sophie didn't know what was happening to her, and was beginning to panic — something was choking her but she didn't know who, what or how. When she turned and saw Steve's hands around her neck, she quite reasonably concluded that he was trying to strangle her. No sooner had he removed the offending item than she went berserk, screaming and trying to bite him.

Steve knew what was good for him, and ran — with Sophie in hot pursuit. I intervened in the nick of time just as Sophie grabbed hold of his leg. I held on to her legs while she continued to hold on to Steve, refusing to let go. She was hysterically and desperately trying to wriggle free of me so that she could bite him. In all the fracas, my finger was bitten and bled profusely. But it had been an accident and Sophie wasn't even aware she had injured me. I managed to prise her away from Steve's leg and took her outside to calm her down. But each time I attempted to go back into the kitchen, she wanted to break free of my grasp and attack him again.

At one point I was holding on to Sophie with all my might and she was clinging to the open kitchen door with her feet. She was less than two years old but it had taken all my efforts to stop her from harming Steve. So determined was she to enact her revenge, that poor old Steve was forced to leave the building for the rest of the morning and miss his tea break.

But I don't want to paint a picture of Sophie as an aggressive individual — she most definitely wasn't. In fact she was very gentle. It was just that, if provoked, she could look after herself. She was quite capable of hurting someone, as my finger can testify, but would always maintain some self-control, unless she felt you were trying to harm her or her family. However, she could become irritable if she was tired and unable to sleep. Usually I would put her to sleep around eight in the evening. But occasionally she had to remind me that she was tired and wanted to go to bed. Then she would look directly at me and emit a two-syllable grunt: 'Aah-ooh.' The intonation was very different from anything I had heard before from any other chimp and its meaning was specific to bedtime. I would then carry her upstairs and tuck her in bed. It has been suggested that chimps have the ability to invent their own personal sounds, using particular grunts to mean certain things. Kanzi, the bonobo, has made up his own different sounds for banana, juice, grapes and 'yes'. I believe Sophie's use of a particular grunt to mean bedtime was evidence of a chimpanzee's ability to use what are called 'protowords'. I also believe that, had I been able to imitate this sound, we would have been taking our first steps towards creating a common rudimentary language.

Putting Sophie to sleep wasn't that simple. When she was a small baby, it had been relatively easy. But as she grew older and more independent, she didn't want to sleep alone, even though she could hear us downstairs and knew we were close by. She wanted me to sleep with her, but her bedtime was, of course, too early for either me or Audrey. So I would be forced to lie with her until she fell asleep. If I attempted to sneak downstairs, she would open her eyes and grab hold of me tightly, and, as the incident with Steve demonstrated, she was already quite strong. This process happened every day for the next year and a half. We never put Sophie and Oliver to bed together on their own in case an accident occurred, such as Sophie poking Oliver in the eye. Strangely enough, we had to do the same with Oliver each night. He would fail to sleep unless either Audrey or I lay with him for at least half an hour.

Oliver would normally go to bed about an hour after Sophie, and would sleep in our bed, tucked up in the middle. Sophie would always insist on trying to sleep on my lumpy chest. This was as uncomfortable for me as I'm sure it was for her, so each night as she grew sleepy I would move her gently over to my side and she would fall asleep. Fortunately, having Sophie and Oliver in bed with us didn't disrupt our own sleep, as they would wake up quite late each morning.

On the day of Oliver's christening we invited friends and family to our home and ended up chatting until late in the evening. Sophie announced that she wanted to go to bed, but I ignored her. I could see that she was becoming short-tempered with the guests, so I told our friends that I had to put the babies to bed. Fortunately, our guests took the hint and decided to go home. So we tucked ourselves up in bed, with Sophie on my left, against the wall, and Oliver in between me and Audrey. It was, I hasten to add, a large bed. Sophie, very tired after a hard day's play, was trying to get to sleep. But Oliver was still very excited and kept chattering away, refusing to sleep and generally disturbing everyone.

Eventually Sophie had had enough. She sat up with her eyes half open and the hair on her head all dishevelled, leaned across my body and, giving out a loud bark, smacked Oliver on the chest with her hand before I could stop her. She hadn't tried to hurt him and it was only a warning, but Oliver was very upset — more emotionally hurt than anything else — and burst out crying. That was the one and only time Sophie ever got angry with Oliver. She was telling him, 'Be quiet, I want to sleep.' I scolded her and eventually we all settled down to some overdue sleep.

In the wild, the night is potentially dangerous for a chimpanzee. After man, the chimp's main enemy is the leopard, which hunts after dark, when the chimp is at its most vulnerable. The infant learns from its mother how to minimise the risks of predation. For about the first five years it sleeps with its mother in the nest. During this time it learns how, where and at what height in the trees to build its night nest. When the infant is about a year old, it begins to make its own attempts at nest building. These clumsy efforts gradually improve with age.

This nest-building trait is unusual in animals. Apes differ from other primates by constructing a new nest each night. These nests are not used for breeding or as a permanent home base: their main function is for sleeping. This nesting behaviour distinguishes apes from other nest-building animals, such as birds, prosimians (for example, lemurs and bushbabies) and many nocturnal mammals, including hyenas. As well as making night nests in trees, chimpanzees often make simple ground nests for sleeping in during the day.

From a very early age Sophie would wrap clothes or blankets around herself to make a rudimentary nest. She would only do this when sitting on the bed. This process was never taught or initiated by me and she never observed it from other chimpanzees. It was an innate action: she was building a ground nest. Sophie, however, never attempted to make a nest in the trees — I suppose because I never taught her how.

Why chimpanzees make ground nests for use during the day is not entirely clear. Not all chimpanzees make these nests, some preferring to sleep on the ground itself or to construct day nests in the trees. Ground nests are simple constructions, normally formed from the surrounding ground vegetation of grass, herbs, shrubs and so on. The chimp sits on the ground and wraps and twists the vegetation around itself to form a ring. This ground-nesting behaviour is innate but probably serves very little purpose today and is not necessary to a chimpanzee's survival. Perhaps it points to a time during chimpanzee evolutionary history when the ancestors of modern-day chimps lived a more terrestrial lifestyle than they do now, similar to that of modern-day gorillas.

This is not such a wild speculation. For it has been suggested that upright walking may have already evolved before the ancestors of humans and the African apes had diverged from a common lineage, and that modern-day chimps and gorillas had ancestors that walked upright and then later evolved once again to adopt a more four-footed gait. The fossil remains of

Australopithecus afarensis — generally accepted as the oldest hominid — have been dated as being around 3.8 million years old. *A. afarensis* — the most famous of which was christened Lucy — was an upright-walking, ape-like hominid, three to four feet tall. The bones of the foot, notably the heel bone, indicate that it walked upright but that, owing to the presence of an opposable big toe, it was also adapted to climbing trees. In 1997 molecular biologists claimed that chimps and humans had diverged from a common lineage some 3.6 — 4 million years ago. If they are right, then the overlap is intriguing, as it suggests that *A. afarensis* was the common ancestor of modern-day humans and African apes.

⋆ ⋆ ⋆

To prepare Sophie for her eventual introduction into chimp society, I tried as best as I could to keep her upbringing as chimp-like as possible. I knew there was no easy way around her introduction to the chimps at Chester and this was destined to be a crisis point in her life — if not mine too. It was going to hit her hard and I attempted to lessen the blow. I groomed Sophie from an early age by parting her hair and pretending to clean her and look for anything that shouldn't be there. Grooming is an important feature of primate behaviour in which individuals carefully examine and clean their own or a partner's body.

Grooming takes two forms: self-grooming and

social grooming. Self-grooming serves mainly a hygienic function, keeping the body free from dirt, dead skin and ectoparasites, and is believed to be innate. Social grooming serves many functions. As well as for cleaning the body, it is also used in a number of social activities, such as relieving stress and establishing and maintaining harmonious relationships between members of the community. Unlike self-grooming, social grooming is learned and perfected over time.

The purpose of grooming in primates changes during the animal's development. When the infant is young, the mother grooms it, which helps both to keep it clean and to bond their relationship. As the infant develops it begins to groom its mother. Also, by observing her actions, it learns how to use grooming as a tool for manipulating others. As an adult, social grooming becomes an essential part of many complex behavioural strategies. Chimpanzees appear to 'trade' grooming for certain favours such as copulation, access to infants and support during aggressive encounters.

In order to be a successful member of a community, a chimpanzee needs to have the backing of friends and family. These social relationships must be regularly serviced and maintained. This is usually achieved through social grooming — you scratch my back and I'll scratch yours. Sophie would need to be proficient at social grooming if later she was to establish bonds and friendships with the chimps at Chester.

If humans and chimpanzees are so closely

related, why don't we groom each other, like other primates? Well, apparently we do, and have largely replaced the subtleties of physical social grooming with a vocal equivalent. It is believed that because humans live in large groups, effective grooming became inefficient as a means for servicing partners and maintaining social bonds. However, by gossiping — essentially vocal grooming — humans were able to maintain social ties with several people at the same time.

As Sophie grew older she would attempt to groom me in return. If I had a scab on my arm she would try to groom it off — a painful exercise that usually resulted in a small scab becoming a large gaping wound. She would grab my head, pull it down towards her and then go through my hair roughly, looking for anything that shouldn't be there. At other times she would bend my nose backwards, quite painfully for me, to inspect inside my nostrils. When grooming her, I would try to imitate a wild chimp's lip-smacking noise, a noise that a chimp makes to its partner to initiate or to maintain a grooming session. If the groomer finds a tick or something on its partner, it becomes very excited and its lip smacks increase in volume. This sound is a reassuring sign to the individual being groomed that their partner is efficient and that their grooming session is proving worthwhile.

Meanwhile, back at Chester Zoo, Sophie's mother, Mandy, who was pregnant again, had been vomiting in the mornings and on two occasions had fainted on the island. This is quite unusual in a chimpanzee. When she fainted she was immediately attacked by some of the other females, notably Halfpenny, as she was unable to defend herself. Chimpanzees are highly skilled opportunists and will often exploit the vulnerability of others. Why Halfpenny did this was not clear, but fortunately Mandy was unharmed by the attacks and we managed to get her safely inside. The vet discovered that her blood sugar was low and that she was slightly diabetic. So it was decided to keep her inside for the remainder of her pregnancy. It was almost a year to the day since Mandy had given birth to Sophie, and she was suffering from morning sickness.

On 25 October 1991 we arrived early in the morning to find that Mandy had given birth to a healthy baby. She was only 14 years old, but this was her fifth baby. Including Sophie, the previous four had all been rejected, and only two of these had survived to be hand-reared. However, this time, to our relief, Mandy had listened to her instincts and decided to hold the baby, albeit in a somewhat clumsy fashion. As soon as she had taken these first steps of motherhood, a bond was able to develop between mother and baby. The baby was named Enya. Sophie had a sister, although she would never be aware of this. Enya was pretty, but looked quite different from Sophie, and it is impossible to know whether they had the same father. We checked back through the record books and found that one of the males, Wilson, had been recorded mating Mandy around seven and a half months before — the normal gestation period for a chimpanzee. Wilson was the beta male (ranked number two) in the group; like Mandy, he too had been hand-reared. He was rarely seen having any sexual interaction with any other chimp. However, on the few occasions when he had been seen mating, it was usually with Mandy. She was his favourite female, it seemed, and the rest just didn't interest him.

Time was moving on and I was constantly trying to prepare myself for the day when Sophie would have to leave me and return to the zoo. It was like a death sentence hanging over me. I would always try to look at it in a positive way, believing that if she had to have a captive life, then there could be no better place than Chester Zoo, home to a collection of chimpanzees surpassed in size by only one other in Europe. At the same time, because I was working there, I would be able to see her each day and keep a watchful eye on her.

When Sophie was about 18 months old I attempted to get her used to sleeping on her own — something she would have to do at Chester. In doing so I was attempting to wean her off me. In the wild, a nursing infant can be a burden on the mother. The mother needs to find a balance between the amount of parental investment she is prepared to give her infant and the needs of both herself and any future offspring she may have. Infants try to resist this weaning process, and the result is a conflict of interest between the mother and the infant. The mother's rejection during weaning is important, as it allows the infant to gradually become independent of her. Although this can be a traumatic period, infants deprived of this weaning process owing to the loss of their mother may not be mentally prepared to cope with their sudden independence.

So, each evening, I would place her in a single bed in our spare room. Then I would lie with her, as usual, until she fell asleep. We had two beds in this room, and I would sleep in the other bed, beside her. At least that was the plan. My intention was to sleep a few weeks with her in the same room and then eventually go back to sleeping with Audrey and Oliver in our bedroom. But Sophie was having none of it, and if she awoke to find I wasn't with her, she would immediately get out of her bed and climb into mine. She did the same each day for the next month. I tried putting her down to sleep alone in the bed and then going off to sleep in the other room with Audrey. But she resisted this even more. Each night she would wake up whimpering, open our door and clamber across our bed to lie on my chest, holding me as tightly as she possibly could. I could see that this wasn't going to work, and eventually decided to do the weaning off at the zoo. So Sophie continued sleeping happily with the three of us.

Part Two

Orphaned

5

Life in Captivity

Despite our being resigned to the fact that Sophie must eventually return to Chester, it came as a massive blow to us all when the day came for her to leave. She was 20 months old when the moment we had always dreaded finally arrived. I had no choice in this matter and at the time there was no alternative. So I was forced to gamble on her safety and even her life, for there was no way of knowing whether her integration would be successful or not. I felt I owed this to her, for as much as I would have liked to keep her as part of my family for the rest of her life, this would have been unrealistic and perhaps selfish: after all, she was a chimpanzee, not a human, and needed to have as much of a natural life as was possible — among her own kind. I always knew that she would have to suffer a little before being accepted, but that her pain and ours would be worth it in the long run. Besides, Sophie did not belong to me. She belonged to the zoo.

Sophie had been a part of my life for almost two years, stuck to the left side of my body like a fifth limb. She spent her last evening at home with us at Meadowtown on 29 April 1992. For both Audrey and me, it was a very sad time.

Sophie slept in our bed for the final time, cuddled up to Audrey, who believed she would probably never see her again. Sophie fell asleep quickly, secure in the comfort she was receiving. For us the night went faster than usual and before we knew it the birds were announcing a new dawn. Audrey got up with me in the morning to spend the last few moments with Sophie. She prepared her bottle of milk and her breakfast of mashed banana and Farley's rusks, and then fed her. Then it was time for me to go to work.

A tearful Audrey gave Sophie a kiss and a big hug, and then I carried her to the car and drove her away for the last time. For Sophie, this was just another day visiting her hairy cousins at the zoo and she was oblivious of the drastic changes happening to her life. I went about my work as normal and wasn't looking forward to going home. I stayed much later than usual after work, playing with her in the kitchen. I could see that she was becoming tired and was ready for her usual nap on the drive home. Neil and I decided to let her sleep in a crate in the kitchen for the first few weeks. She was used to the crate and would happily go in and out of her own free will. So, that evening, I placed a couple of bananas inside and she jumped straight in. But, as soon as her back was turned, I quickly closed the kitchen door and locked it. She screamed and made a dash for the door but it was too late. I hurried out of the building and headed towards the car park, her screams fading into the background. As well as grief, I was overcome with a terrible sense of guilt that I had let her down and ultimately betrayed her trust in me as a parent. I drowned my sorrows with a couple of beers in the local pub and then drove the long journey home. I found Audrey in the kitchen and tried to put on a brave face.

'Well, she's gone now, but she'll be OK,' I said, trying to reassure her. But my voice began faltering before I could finish the sentence and we both burst into tears. Without Sophie, the house was silent and empty.

Sophie's life had gone full circle and she was now back at her place of birth. During her early years at Meadowtown she had grown into a confident young infant, benefiting from the security of being loved and cared for. She was a big, healthy baby, almost a head bigger than Kaylie, who was born the day before her, and Kaylie was herself quite a large baby. Her serene and settled life was about to be turned upside down. At Chester she would be thrown into another world — a world of chaos and volatility, of calmness one minute and brutality the next, a world where adrenalin pumped through the veins of each member of the captive community.

Throughout her life Sophie had been surrounded by people who had always shown her affection. Now she needed to be conditioned for the shock of coming face to face with unfriendly, uncivilised individuals who could harm her. She continued sleeping in her crate in the kitchen over the next three weeks. After that we began getting her used to playing inside one of the spare chimp beds. What we called a 'bed' was in fact a large room about 16 by 13 by 10 feet high, and this was where her introductions would take place. We moved her crate into this room, and on 22 May she slept here for the first time. Sophie refused to let go of my waist and clung on with her arms and legs, so strongly that I couldn't get her off me to put her down. In the end I managed this by taking off my shirt and giving it to her as a comfort blanket. She immediately relaxed her grip and begrudgingly allowed me to place her inside the crate. She sniffed my shirt and clutched on to it tightly. As the door closed on her, she whimpered desperately at me and was clearly very upset. This time she was sleeping not far from all the other chimpanzees, who were on the other side of the corridor, and she was very scared. They were used to seeing her with me, and were curious to see this small, disconsolate figure all on her own.

I arrived at work early the next morning and was disappointed to find the outside of Sophie's crate plastered with faeces. After I had left, the other chimps had pelted her with shit from across the corridor. Sophie wasn't used to any form of aggression, so I could only imagine how frightened she must have been all on her own. All this must have been totally incomprehensible to her. I let her out of her bed, took her into the kitchen and cleaned her hair. Chimps are fastidiously clean animals, and generally hate treading in faeces or getting it on them. They also know that others hate it too, and so one of the best ways to annoy someone you don't like is to throw shit at them. Over the next few days the other chimps gradually became used to her presence and the throwing gradually decreased.

Each evening I gave her my shirt or the T-shirt I was wearing underneath it. By the end of the day it carried my body odour — something that she identified with and that gave her some small comfort, much as human children gain reassurance from a comfort blanket. Human and chimpanzee infants often try to make up for the absence of an attachment figure by wrapping their arms tightly around their own body or by clinging on to some favoured object. The other chimps received woodwool — made from shredded wood, this is similar to straw but less likely than straw to carry mould spores — and also hessian sacks. These were spread around the chimp building, so that as the chimps came in from the island they would grab a wad of bedding and select their own sacks before settling down to make a nest for the night.

On the evening of 22 June I put Sophie in the crate for the last time, as from then on I was going to keep myself away from her during her introductions. This had been the zoo's normal practice when hand-rearing chimps. It was felt that, with me around, Sophie wouldn't make any effort to bond with another female. Instead Neil and Steve were going to introduce her to the rest of the group. With 25 members in the group, it promised to be a long and potentially painful process, and I had no way of knowing if I would ever see Sophie alive again. I stayed away from her for eight weeks, foolishly thinking she would socialise better without me.

Each day the staff would keep me informed as to how Sophie was coping. Then, in the evening, I would pass on the news to Audrey. But, at home, I was mourning her. Everywhere I looked, reminders of Sophie stared me in the face. I would go upstairs alone and gaze at the bed and all her photos on the wall, and tears would fall. Oliver was also subdued. He no longer had the playmate he had known since birth. He had never known a day without her and now he played alone. He was only 14 months old, too young to understand, but we tried to explain that she had gone back to the zoo to be with friends of her own kind. It must have been very hard for Audrey to cope with the sudden loss of Sophie. Unlike me, she wasn't able to travel to work each day and find out how she was getting on. She had to rely on me giving her information when I returned home in the evenings. But at least she was helped by having Oliver to occupy her. In the end it was probably better that she was spared the distress of witnessing Sophie at the zoo. I always felt it was like visiting your child in some third-world jail, and it left you with a feeling of helplessness and permanently depressed.

For the next few days I worked only with the orang-utans. All along I was praying for Sophie's safety and couldn't bear to be around to hear her suffering. I dreaded answering the phone in case it was Steve or Neil with some bad news. I suppose it was just as well that I wasn't around, as it would be worrying for me, distracting for Neil and Steve and even worse for Sophie. She was going to have to make a big effort at socialising.

The introductions needed to be done relatively swiftly, to avoid disrupting the cohesion of the group. They took place inside the building, while the other chimps were on their island. The chimps loved to spend their days outside. There were trees to climb, ropes to swing from and even a hill with a tunnel. The island was surrounded by a moat filled with water. The chimps would lie in the sun or fish with sticks for morsels of food from the water, such as floating crisps or bread, or simply pass the time by watching the humans watching them. Not all the chimps would fish: this fishing skill tended to be handed down from mother to infant, so that those mothers who were proficient at fishing usually had infants who were the same.

We decided that Sophie's first introduction should be to her mother, Mandy, not because they were related — neither knew this — but because Mandy was a friendly chimp who we were confident wouldn't harm her. Mandy was also caring for little Enya, Sophie's sister, and so in theory should be preoccupied with maternal duties. When Mandy first walked into the cage carrying her baby, Sophie's first reaction was to literally shit herself. She then tried to get out of the way by climbing up the bars to the top of the cage. After half an hour or so she eventually came down to the floor and sat by the door. She and Mandy were kept together for an hour and then separated. Throughout this period they simply ignored each other. Mandy spent most of the hour grooming her baby and Sophie just sat the entire time by the door, waiting for me to take her away from this uncultured animal.

The next day we increased their time together to four hours. Although Sophie was a little less scared than the previous day, there was still no interaction. During their time together Mandy showed no recognition of Sophie and, of course, Sophie knew only me as her mother. Besides, Mandy was preoccupied with rearing her baby. Sophie was now alone. I was no longer around to intervene if things got out of hand, and only fate would determine whether she would come through unharmed. Every day would become a battle for survival, a test of her social skills and her determination.

I can't pretend that I believe in reincarnation, but given the choice of coming back as another creature, I can safely say that a chimpanzee would be low on my list — not for me their life of chaos and volatility. In the highly conflictual world of chimpanzee society, barely a day passes without some stressful interaction, especially in captivity, where the opportunities to disperse are denied. Chimps live in a dominant hierarchical society, where the laws of the jungle prevail: bully or be bullied; dominate or be dominated; and if you can't beat them, join them.

We kept Sophie with Mandy and Enya for a month, before introducing her to any other chimps. For her next introduction we chose the oldest female in the group, Meg. Over 40 years old, Meg was Mandy's mother. She had abandoned Mandy at birth, and so neither of them knew they were related. Meg was, of course, Sophie's grandmother. So, unbeknown to the four of them, they were all closely related. By now Sophie had grown used to Mandy, and although she was a little wary at first, she wasn't too bothered by old Meg. Despite the fact that Meg paid almost no attention to her, it was good for Sophie to at least get used to adult chimps, as she would soon be meeting many more — much bigger and more aggressive individuals.

Despite having been born at the zoo, Sophie had no friends or allies in this chimpanzee world. I wasn't a member of this group, and so had no chimp allies of my own. Previously, she had spent only the first day of her life there, abandoned on the floor. Infant chimps tend to initiate interactions with their mother's 'friends' more readily than with other adults. Infant and juvenile chimpanzees have the ability to recognise their mother's relationship to various individuals and are able to adjust their behaviour accordingly. In human infants this ability is called 'social referencing'.

Sophie's period of separation was proving highly traumatic for her. She became depressed and it wasn't long before she began demonstrating classic symptoms of maternal deprivation — stereotypic rocking and self-clutching behaviour. She would sit with her back to the wall and rock backwards and forwards so that her back banged against the wall with each movement. At the same time she would constantly whimper. As her back hit the wall, her whimpers became broken into a pathetic staccato sound: 'Aah

. . . aah . . . aah . . . aah . . . aah.' It was very upsetting, and even though I couldn't see her, I could clearly hear her suffering. She was behaving like an abandoned child and, ironically, had become what in essence she had always been since birth — an orphan.

There have been many studies on the effects of maternal deprivation in non-human primates and in humans. It was found that in infant rhesus macaque monkeys, aged between 18 and 32 weeks, a temporary separation from the mother of just a week could still be detected in the behavioural responses of the infant up to two years later. Most notably, these infants showed a greater fear of strange objects. Animals that were reared in complete isolation tended to have difficulties interacting with other adults and were often inefficient mothers, failing to rear their own young.

Isolated chimpanzees frequently console themselves with actions such as rocking and self-clutching, as Sophie did. The loss of the mother often results in the orphan becoming depressed, insecure and more vulnerable to aggression. Males, especially, have fewer friends, suffer more harassment and receive less support during aggressive encounters. Their social and sexual behaviour can be affected, and sometimes this leads to their being sexually retarded. In a study of 71 chimpanzees reared in isolation with human care-givers, only 30 per cent became sexually competent as adults. Wilson, like Mandy, had been hand-reared, and I believe his inadequate social skills were manifested in his lack of sexual activity.

Each evening, as the rest of the chimps came inside from the island, they would immediately check on what was happening with the separated members of their group. Many had previously been through several introductions during their lives, so they all knew the score. They knew what was happening and that shortly it would be their turn to be parted from the group and introduced to Sophie, whom they associated with me. Not surprisingly, the chimps took exception to being kept indoors and resented Sophie's presence, as they correctly saw her as the reason for their incarceration. The males especially resented being separated from their females.

Now that Sophie was successfully integrated with three chimps, the introductions proceeded at a faster pace. Towards the end of July we introduced Nicky. He was a huge, muscular male, the biggest chimpanzee I have ever seen. He was also a terrible coward and known among the staff as the gentle giant who wouldn't hurt a fly. Nicky entered Sophie's cage and initially took little notice of her. When she saw this monster of a chimp coming in she screamed, leapt about four foot into the air, and scrambled to the top of the cage as fast as she could. Nicky tried to calm her by adopting a submissive posture, bending over and presenting his rear to her. They were kept together for about an hour and then separated. This procedure continued each day as we slowly increased the time they spent together.

Sophie seemed to be gradually establishing an

uneasy relationship with Nicky. But then one day, just before they received their food, we heard her cry out loudly. Steve rushed to check on her and found her sitting on the floor by the door, rocking backwards and forwards frantically. There was blood in her mouth. Nicky must have taken a swipe at her and hit her in the face. We immediately separated them.

We kept Sophie with Meg, Mandy and Enya for a few days. Then we introduced another infant female chimp, Sally, who was the daughter of Rosie. Although Sally was a youngster herself, at about four years old, we had to be careful when introducing juveniles as, like human children, they can be prone to bullying. However, Sally was too upset about being separated from her mother to bother with socialising. The next morning I was informed that Sophie had a small cut on her left ear but that it was apparently nothing for me to worry about. It worried me a great deal, of course.

In fact Sally was feeling as miserable as Sophie, and as a result wasn't interested in making friends. After two days of Sally's howling for her mother, Neil and Steve could take it no more, so they introduced Rosie to Sophie. Things seemed to go quite well and Sally, at least, was relieved. But, the following day, Sophie had a chipped front tooth. How or who, nobody knew.

Ten days later Steve gave me some encouraging news. He had seen Sophie playing with Sally for a few brief moments, and even witnessed them embracing in the evening. But, strangely enough, Meg had been unhappy with Sally's interacting with Sophie in this way and had pulled them apart. She then began prodding and bullying Sophie. Meg's show of discrimination — not wanting the infants to interact with this outsider — seemed human. Maybe Meg had been harassing Sophie all along.

A week later we introduced another young adolescent female, Sarah, who was about a year older than Sally and the oldest daughter of Halfpenny. The introduction took place with little incident. In fact there was no interaction at all between Sophie and any of them. All the chimps knew who Sophie was — they had seen her with me for over a year. What they made of this, and whether they thought she was my infant or that I was looking after a member of their group, is impossible to tell. But the chimps always reacted differently to visitors carrying children from the way they behaved towards people carrying cuddly toy apes or teddy bears, becoming aggressive and throwing sticks or stones at the latter.

In the evening, while Steve was giving Sophie her milk, he noticed that again there was blood in her mouth and that she had lost a front tooth — not the one that was already chipped but another one. Sophie was separated once more, although we couldn't keep her alone for too long otherwise we would have to go through the whole process of introducing her to the same chimps all over again, which might be dangerous. I tried to hide my feelings from my work colleagues. But inside I was suffering over

Sophie, and each injury she received only increased my own pain.

A week later I was informed that Sophie's chipped tooth was further damaged, and then, the next day, broken even more. Her mouth and teeth were bearing the brunt of the attacks on her. At this rate she will have no teeth left at all, I thought to myself. We called in the zoo vet Derek Lyons (yes, Lyons really was his name), who sedated her. He also took the opportunity to weigh her — she now weighed 20 pounds (9 kilograms). I don't think Sophie was aware that I was with her during this time. It was very upsetting seeing her lying prostrate on the table. Poor Sophie was going through a nightmare and there was no easy way through this.

Throughout this process I had been hoping that she would forget about me and make friends with her own kind. So I was horrified to find that, rather than interact with any of the others, she simply sat by the door and just waited for me to come back. In fact she waited every minute of every single day of those eight weeks. This was probably the reason why her teeth were taking such a pounding. It may be that Sophie was being hit on her back or on her head, which caused her face to smash against the bars. I would continually ask my colleagues how she was coping, and they would try to reassure me that she was OK. In the end I decided to secretly check on her myself, so I found a large piece of cardboard and cut out two eye holes. With this mask on, I stood at the other end of the corridor, nearly 40 feet away from where she was sitting.

But as soon as I put my head around the corner Sophie leapt up on to the bars, whimpering with her arms stretched out towards me pleading for me to come to her aid. All she could have seen was a pair of eyes in the distance through the cardboard. My heart sank, for I knew instantly from her reaction that she had somehow known it was me, and I had again failed her.

The plan wasn't working out well at all and Sophie and the others continued to take little notice of each other. So, towards the end of August, after a separation of eight weeks, I finally gave up on our strategy and went over to Sophie. She had been waiting two months for this moment and immediately greeted me enthusiastically, with her arms held out, expecting me to open the door and rescue her from her nightmare. This was my only wish too, but it was something I couldn't do. So I sat down on the floor next to her. I could see she'd been having a rough time. Her mouth was swollen, her upper lip cut and both ears were puffed up and scratched. She looked relieved to see me at last.

Surprisingly, Sally was also pleased to see me, and she walked over and hugged Sophie. Within five minutes the two of them were playing together. This was the first genuine interaction between them and it seemed as if Sally could sense Sophie's happiness and was pleased for her. It was the first time Sophie had had friendly contact with another chimp and it was reassuring to watch them. She played non-stop with Sally for one and a half hours. She even played a little with Meg and Rosie.

Play makes an important contribution to the social development of an infant. It allows the individual to gain personal knowledge of other group members; permits certain social skills to be practised, such as fighting, without risk of injury; and helps the individual to establish long-term friendships.

I was relieved, and stayed several hours with her until it was dark outside and time for me to go. By this time Sophie was exhausted after her exertions with Sally and lay down close to me by the door. I whispered a vow in her ear that I would somehow get her out of here. She seemed so happy to be with me, but when I got up to leave she screamed and whimpered louder than ever before. I left hurriedly, my head filled yet again with feelings of sadness and betrayal.

Back at home I told Audrey the encouraging news about Sophie and the time I had spent with her. I played down her injuries so as not to make her any more worried than she already was. I had been reassured a little after seeing the way Sophie was willing to play with Sally in my presence. I said that maybe things would go more smoothly now and Sophie would begin to settle in with the others.

The next morning I went back and sat with Sophie again. Although she looked very subdued, she was clearly glad to see me. She seemed less interested in playing with Sally, maybe because she realised that I wasn't taking her home. I had hoped that their friendship would go from strength to strength, but I could see that their initial enthusiasm had waned.

The weekend came, and we thought we would cheer ourselves up with a visit to the Welsh border town of Hay-on-Wye, famous for having the most second-hand bookshops in Europe. I enjoyed collecting books on wildlife and the environment and Audrey had a passion for art. If you can't find the book you're looking for in Hay-on-Wye, then you probably won't find it anywhere. On the way back we stopped off at a restaurant for tea and scones. As we were paying, Audrey started browsing through the tourist pamphlets by the till. One of them caught her eye. It was for the Welsh Mountain Zoo at Colwyn Bay and she was shocked to see what appeared to be a photo of Sophie on the front cover. She called me over and I felt my heart skip a beat as I looked at what indeed seemed to be Sophie. I agreed that there was a remarkable likeness. The eyes, hair and overall features were Sophie's. But we knew it couldn't be her. However, I had a hunch that I might just know who it was. So the next day I rang the Welsh Mountain Zoo and asked them the name of the chimp on the pamphlet.

As I suspected, the chimp was called Tuppence. I had never seen a photo of this chimp before, but I did recall that, a few years earlier, Mandy's only other surviving baby had been hand-reared and sent to Colwyn Bay. That infant was called Tuppence and she was Sophie's older sister. This seemed too much of a coincidence. We were trying to get Sophie out of our minds, only to bump into a photo of her unknown sister in the middle of the Welsh countryside!

I returned to the zoo the next morning and found Sophie waiting for me at the door. I gave the other chimps their breakfast and then arranged for the vet to anaesthetise Sophie to check on the state of her mouth, which was still badly swollen. Sadly, it proved necessary to extract three of her teeth: one top-left upper incisor and two bottom-right incisors. She was then put on a course of antibiotics for a week.

Over the next few days things seemed to calm down a little, so again we introduced Nicky to Sophie. She seemed less nervous of him now, although this may have been because I was with her and she felt more secure. Even so, she remained anxious with him around, and kept a constant check on his whereabouts.

Things were progressing reasonably well, and Sophie was beginning to gain confidence and getting used to being with the largest of all the chimps in the group. I was hoping that Nicky would be able to offer her the protection that I couldn't. One evening while I was feeding all the chimps, I noticed that he had a tooth that was badly cracked very close to its base. We called in the vet, who decided to anaesthetise him in order to make a closer examination. Because Nicky was going to be anaesthetised the next day we were unable to feed him that evening. The next morning he was given milk but no food. When he saw the other chimps being fed he became very frustrated and angry, and started prancing up and down his cage with an irate expression on his face. It was obvious he was annoyed with me.

Nicky wasn't normally the type to show any aggression towards the staff, so he decided to take his frustration out on the next best thing — Sophie. This was either because he obviously associated her with all his problems or because he wanted to get back at me. Without any warning, he walked up behind her and attacked her in front of me. Grabbing her by an ankle, he shook her around violently as if she were a rag doll. How he didn't break her leg or seriously damage her ankle ligaments, I will never know. He then pinned her to the ground and started biting her head, one of his massive canine teeth penetrating the top. I was helpless and could only scream at him to stop. I grabbed a brush and tried to push him off with the handle, but in vain.

Eventually he let her go and sat on one of the platforms watching me. Sophie, battered and bemused, crawled into the corner behind the door. She sat there whimpering and rocking, with her back to the wall. Instinctively, I opened the door and stepped smartly inside. I put my arm around the back of the door, grabbed her arm and pulled her out. All through this, Nicky just sat there watching me. With his immense speed and strength, he could have easily leapt at me and I would have been powerless to protect myself. As I carried Sophie away to the staff quarters I had visions of her having sustained brain damage. We called the vet. Nicky's canine had left a deep hole running parallel to her skull beneath her scalp. It had scarred her skull but, to my immense relief, hadn't penetrated the bone. Sophie's ankle was also damaged, and each time she moved her foot you could hear the ligaments grind together.

This was a massive setback to Sophie's self-confidence and to our integration programme as a whole. It left me feeling much more insecure about her long-term survival prospects. If Nicky, the gentle giant, could do this to her, then what would happen when she was eventually introduced to some of the more dominant and aggressive males and females? A small piece of bone, a couple of millimetres in length, grew over the wound, leaving a small lump beneath her scalp. Her injuries eventually healed, leaving a large, round scar on her head, but her shattered confidence took much longer to recover.

We kept her separated from the others while she was recovering, and she was put on a course of amoxycillin. She was relieved to be away from the chimps and back with me and her other human friends. She knew humans weren't like chimps: they would cuddle and play with her and never harm her. We let her run around in the staff quarters for a few days. In the kitchen there was a small window that looked out on to a public footpath, and hundreds of visitors to the zoo would walk past it every day. One day Audrey's mother was visiting me at work. A keen gardener, Solange was more fond of the zoo's impressive flowers than of the animals, and happened to walk past the window. Sophie couldn't have seen her for more than a split second but her reaction was spontaneous and dramatic. In a flash she jumped down from a cupboard and ran whimpering to the window. With her nose pressed firmly against the glass, she stared outside in the direction that Solange had walked. She stayed there for some minutes, pouting and whimpering. I was amazed by her memory. She had met Solange only a few times and hadn't seen her since last Christmas, more than eight months earlier. I don't believe that a human would have reacted so quickly and so assuredly.

During the time that Sophie was separated, we had a severe outbreak of flu in the group. Many of the chimps became seriously ill. Chimps are susceptible to almost all human diseases, and vice versa. In fact I have caught many a cold from the chimps at Chester. Whenever there was a flu outbreak in England we would all wear masks to try to minimise the chances of infecting any of the chimps. We were especially concerned for the infant chimps as flu could be a killer. On this occasion at least half a dozen chimps became very ill and were put on antibiotics. The problem was that it was near impossible to give the very young babies any medicine as the mothers wouldn't allow it.

One female, Florin, had a young baby about a year old called Alice, who was ill. I arrived at the chimp house one morning and turned on all the lights, to see Florin waiting for me at the bars. Florin was Halfpenny's sister and, just like her, was very unsociable when it came to interacting with the human staff. But today was different, for she tried to get my attention by putting her arm out to me. So I went over to say hello. Then she grabbed hold of little Alice's arm and thrust it through the bars to me and whimpered. She did this twice in quick succession. I was very shocked because Florin normally wouldn't let me anywhere near her baby. Alice was extremely weak, with hardly the strength to lift her arm. I could see that Florin was very distraught and wanted me to help, so I immediately called Derek Lyons to come and check on Alice. He decided to anaesthetise Florin and put the baby on a course of long-acting antibiotics. A further strange thing then happened. All the chimps knew the vet, and associated him with unpleasant experiences such as painful darts thudding into their backsides. Whenever they saw him they tried their hardest to avoid him and when they saw his blowpipe they were all petrified.

However, on this occasion Florin calmly went up to the bars in front of Derek, turned around and presented her side to be darted. This he gratefully did. In fact he didn't need to use his blowpipe, but just injected her by hand with a syringe. Florin then climbed on to the bars and waited there, before falling asleep still holding on to them. We went into her cage and tried to help her down to the floor. Alice was petrified and clung firmly on to her mother. It took all our strength to prise Florin's fingers free from the bars, even though she was asleep. I was amazed by her grip. The baby was successfully treated and made a complete recovery. This episode was powerful proof of a chimpanzee's intelligence and ability to communicate using gestural language.

Most of the chimps in the group at some stage became ill with this flu. Fortunately for Sophie, she was separated from them and so was not affected. But once the outbreak was over, the introduction procedure had to continue — paradoxically, for her own safety. The longer we kept her away from the others the more dangerous it would become to put her back in with the chimps she already knew. But, of course, we separated Nicky from her group. In the middle of September I was pleased to see that Sally allowed Sophie to share some of her food. This was a positive sign; maybe she was beginning to be accepted by at least some members of the group. However, my hopes proved somewhat premature and a week later her top-right incisor number two was broken in half. I kept telling myself that these were only her milk teeth and that if these proved the only injuries, then it was painful but a price worth paying.

Towards the end of the month we decided to introduce Heidi and her five-year-old daughter, Wanda, to Sophie. Heidi was a rather overweight chimp. Her daughter had inherited this trait and was also quite rotund. Wanda initially teased Sophie, but after a short while began playing a little with her. By the same evening Heidi was crouching in front of Sophie in an attempt to encourage her to climb on to her back. Sophie would do so and make some effort to hold on, but after a few seconds would jump off.

Basically, she didn't know what was expected of her. In retrospect, maybe I should have made more effort to carry Sophie on my back and walk around on all fours when she was living with us at home. Nevertheless, these were encouraging first signs, so I left the zoo that evening in relatively good spirits, hoping that maybe Heidi would adopt Sophie and protect her from the others. This is what she really needed and it would make the other chimps accept her more readily.

Several months had now passed since Sophie was first introduced to Mandy and things were going fairly well, if slowly. By now she had been introduced to several of the 25 chimps, but she was still not fully integrating. It was hoped that a female might foster her. But it takes two to tango — the adult must want to foster the infant and the infant must want to be fostered by the adult — and most of the eligible foster mothers already had infants of their own. But, more importantly, Sophie didn't want to be fostered. Why should she, when she already had a parent in me?

The following morning I entered the chimp house and went over to see her. I was hoping to see her playing with Sally or Wanda. But instead she whimpered loudly when she saw me. She was letting me know that she was hurt. The left side of her face was distorted and horribly swollen, like a football. Her left eye was swollen and tightly shut and her nose and mouth were cut. Her right foot was also swollen, and bleeding. On top of that, two of her front teeth (the few remaining ones) had been broken. It was a terrible sight and she must have endured a long and prolonged beating throughout the night. I quickly got her out of the cage and took her into the kitchen. She was relieved to be out of her living hell. We never found out for sure who had attacked her. It could have been Heidi, frustrated with Sophie's lack of response to her advances; or maybe it was Wanda, jealous of the attention Heidi was displaying towards her. Or was it the elderly Meg, angry once more that Sophie was interacting with members of her group?

Poor Sophie was put on amoxycillin again. I was concerned that she would begin developing a resistance to this important antibiotic. We separated her from the others for a week. Then, on 7 October, Sophie was put back in with Rosie and Sally. The following day I introduced Heidi to them and Sophie screamed when she approached her. I felt pretty sure from her reaction who the guilty party was and immediately removed Heidi.

After a week we introduced Gloria to Sophie, Rosie, Sally and Meg. Gloria was a slightly neurotic chimp. She was also a very frustrated female. She had given birth on several occasions but all her babies had been stillborn. Gloria would often carry other chimpanzees' babies on her back. She loved babies and, despite not getting on with Halfpenny, had become Sarah's adopted aunt. Gloria was one of the few eligible females who could potentially adopt Sophie. Things went reasonably well at first, and there was a little interaction between them. But a week later I witnessed Gloria bite Sophie on the finger. Sophie screamed out in pain. Her finger was cut and bleeding. I shouted angrily at Gloria, loudly using my chimp 'uhh!', and chased her around the bed, threatening to throw her milk at her. Feeling braver with me around, Sophie screamed at Gloria and wanted to attack her as well. She was looking over her shoulder at me for reassurance and support, making sure that I was right behind her. Realising that the situation was potentially dangerous for Sophie as I was on the wrong side of the bars to protect her, I knew I had to try to calm her down.

The next day Sophie's ear was bleeding — possibly again bitten by Gloria — and a week later her mouth was once again swollen. She began losing her confidence, which further affected her relationships and social standing with some of the chimps who had already accepted her. Two days later Sally began bullying and tormenting Sophie. She obviously felt that this was more fun than playing with a chimp who in any case spent most of the time ignoring her.

It was all very depressing and I kept asking myself when it would ever end. Sophie had suffered systematic beatings and mental torture. But even though she was so young, somehow she was still coping with it all.

Some days later I received another call from Jane Goodall. She wanted me to return to the Congo for two months to assist in moving 34 orphan chimpanzees from the Brazzaville and

Pointe-Noire zoos to the newly constructed Tchimpounga Sanctuary at M'Pili on the coast. This couldn't have come at a worse time, and in any other situation I would have refused. But secretly I hoped that somewhere in Africa there could be the solution to Sophie's nightmare. I convinced myself that she would manage without my support and accepted the mission. By now it was mid-November, and once again Jane wanted me to go immediately. So, on the evening before my flight, I sat with Sophie for much longer than usual. She lay down by the door next to me and I waited until she began to get sleepy. Then I whispered goodbye and slipped away.

6

Back to the Congo

Graziella was at the airport in the Congo to meet me. She was happy to see me again and also pleased that I had kept my promise and brought with me a high-pressure hose, which Chester Zoo had kindly donated. We loaded it on the back of her pick-up and drove to her house. It was good to be back in Brazzaville again. I was warmly welcomed by Jean Mboto, and pleased to see that he was still here, as dedicated as ever to the cause. Most of the animals were also still alive, including the old male chimp Grégoire. Even the two lions were still there, as skinny as ever. I was amazed to see how well the three young chimps, Akim, Pasy and Jacob, were getting on. Graziella and her staff were obviously doing a great job. Pasy, in particular, seemed transformed from the pathetic, helpless individual I had known just a year ago, and was now big, strong and good-looking.

But, outside of the zoo, things no longer seemed the same, and I was rather disheartened to find that the political climate of the country had greatly changed since my last visit. There was a newly elected president and a new prime minister, and with them new ideas and fresh hopes. The former president Denis Sassou-Ngeusso's Marxist government, drawn from the Congolese Workers' Party (PCT), had recently been replaced in August 1992 by a so-called democratic one, the Pan-African Union for Social Democracy (UPADS), led by Pascal Lissouba.

But this was no smooth transition, as another political figure had come to the fore. Bernard Kolelas, leader of the main opposition party, the Congolese Party for Integral Development and Democracy (MCDDI), had narrowly lost the presidential election to Lissouba and was loudly voicing his dissatisfaction, challenging the fairness of the elections and demanding a rerun. Kolelas had earlier formed a coalition with Lissouba on the understanding that his party would receive a share of the power. However, when pay-back time arrived, his party was offered only three out of 28 cabinet seats. He was also unhappy with Lissouba's choice of prime minister, former military leader Jacques-Joachim Yhombi-Opango. Yhombi-Opango had been imprisoned for 11 years after being overthrown in 1979 by Sassou-Ngeusso, and was now looking to 'get even'. In Africa it is safe to say that politics is truly personal. But the policies remain the same, with only the faces changing. Most politicians in Africa exploit tribal intolerance to gain popularity and power.

I had landed in the middle of a power struggle, and the normally stable Congo was now locked in a civil war, looking more and more like its politically volatile neighbour Zaire.

The country I had not so long ago admired and described to friends as being one of the safest and politically stable in Africa had changed dramatically over a short period of time. There were countless roadblocks, with soldiers and tanks on every corner. For three days the government announced what they called a '*ville morte*', or dead town, and no cars were permitted to be driven in Brazzaville. One morning, not far from the zoo, eight people were killed and a further 27 injured. Although we were advised by the US Embassy to remain indoors, we carried on working as usual at the zoo. Throughout the day there was the crackle of gunfire every 15 minutes or so, and the situation felt very tense.

That evening, as I strolled back towards Graziella's house, a 20-minute walk, I could hear loud cheering. A rally was being held at the nearby football stadium by Kolelas, and the atmosphere was electrifying, as he had whipped up fanatical support. At times the cheering was explosive — it sounded more like the winning goal was being scored at a cup final. As I crossed the busy road leading to the airport, I noticed a Nissan taxi, full of passengers, heading in my direction. Noticing the driver start to accelerate straight towards me, I coolly continued to cross the road, maintaining my same pace. I thought he was bluffing, but he still sped towards me until, at the last minute, I made a desperate leap to the safety of the kerb. The minibus missed my trailing foot by inches. Several laughing faces peered out through the back window as the vehicle disappeared into the distance. I was a bit shaken but tried to put this unpleasant incident behind me as a one-off.

The bush-meat trade was continuing to produce a steady stream of orphaned chimpanzees, bonobos and gorillas, and both of the zoo's orphanages were becoming full. Hunting is now considered to be the main threat to the survival of the great apes. It is feared that, unless urgent action is taken, within 50 years there will be no viable populations of great apes left in the wild. In Yaoundé, the capital of Cameroon, 70–90 tonnes of bush meat were being sold each month in 1998 alone. In fact it is estimated that, throughout Africa, up to one million tonnes are consumed each year.

One sweltering afternoon I stopped for lunch at a local restaurant close to the zoo. I was struggling with my tough and stringy chicken and chips when two men approached me. One was carrying a bag. They said they had seen me working with the chimps at the zoo and wanted to know if I was interested in buying another. To my astonishment, one of them then opened the bag to reveal a very sorry-looking chimp. She was a young female around three years old, and looked like a bag of bones, weak, badly malnourished and dehydrated. Her eyes lacked any co-ordination and seemed unable to focus on me, wandering from side to side all the time. She had a six-inch scar across the top of her head. When I asked the men how she had got this injury they told me, unashamedly, that they had killed the mother but the infant refused to let go of her dead body, so they knocked her out with a machete.

I said I might be interested but that they would have to come back to the zoo the next day because I was unable to make this decision alone and would have to ask my boss. Naively, they agreed to this, and immediately they had left I went to see the wildlife authorities and told them what had happened. Conveniently, their office was in the zoo grounds. They suggested that when I met these men the following day, they would pose as my employers and arrest them. Reluctantly, I agreed, although I felt I was becoming embroiled in some dangerous adventure.

I woke up early the next morning and noticed that a tiny, innocuous-seeming scratch on my right hand from the previous day was now inflamed and itching. Because of the high humidity, the slightest injury could quickly become infected. Overnight my scratch had become a six-inch red streak along one of my veins, from the knuckle to just above my wrist. Fortunately, Dr Cedric Dumont, the US Embassy doctor, was around to diagnose the problem and treat me. He told me that I was coming down with septicaemia and gave me the necessary antibiotics, free of charge. At the time I didn't feel particularly unwell, but he told me that, had I left it another day, I would have become seriously ill.

I arrived at the zoo that morning to be told about a horrendous accident that had happened in my absence. A young French couple had visited the zoo and brought some food for the lions. The husband asked his wife to pose in front of the lions' cage with the meat held out in her hand. Foolishly, he made her stand on the cage side of the safety barrier. One of the lions had managed to grab hold of the woman's arm through the bars and had bitten it off at the shoulder. The woman was rushed to the city hospital. I was shaken by the news, and thankful that I hadn't been around to witness the gruesome event.

Later that morning the men arrived and found me feeding the chimps their breakfast. My heart immediately started beating fast and noisily. I was really nervous and felt sure that they would notice this and get suspicious. But when I shook hands with them they failed to notice my trembling hand. I went and called the men from the wildlife authorities. Pretending to be the zoo's directors, they asked the two men to accompany them to their office, where they revealed their true identities and confiscated the chimp. The outcome of all this was that the men were allowed to go free with just a warning. But, as I feared, they returned the next day and asked one of the zoo staff where I was. I decided to face the music and went to speak to them.

The whole episode was nerve-racking. The men angrily asked me why I had set them up. I played dumb and replied that I didn't know that it was illegal to sell a chimpanzee and that, as I didn't work here and was only helping out, I had no idea what my bosses were intending. Surprisingly, they seemed to accept my lame excuse and left, muttering their discontent. Graziella named the young chimpanzee Rita. The little thing was terribly weak and it was clear, from her wandering eyes, that she had a serious head injury. We gave her rehydration fluids and tiny pieces of mashed banana. But, sadly, despite all our efforts, she died two days later.

Meanwhile things were hotting up on the political front and that evening I went to Mark and Helen Attwater's house and we tuned in to the BBC World Service. An announcement on the radio advised all foreign citizens to leave the country immediately. We were close to being evacuated on the eve of moving the chimps down to Pointe-Noire. I went back to Graziella's and we rang Cedric to discuss what to do. We decided we had to go ahead with the operation the very next day or risk abandoning the plan altogether. So we arranged to rendezvous at the zoo at six the following morning.

We all met up on time and set to work trying to get all the 25 young chimps into five large crates. The crates were quite long and had a sliding divider in the middle. We coaxed each chimp individually inside with food and then quickly closed the door behind it. Then we pushed the chimp to the far end and partitioned the crate, leaving the front half empty. When we had caught another chimp, we would open the divider, put them together and then push them to the far end again, leaving the other half empty. This was easier said than done.

We also had to make sure that we put only

chimps that were friends together in the same crate. Some of the larger chimps were determined not to go inside, especially once they realised that it was a trap. But, after a long struggle, we managed to capture all of them. The whole exercise took almost eight hours, and we were all extremely tired. We kept the chimps in the crates overnight ready for their long trip to the coast the following morning. The three adult males, Grégoire, Zou Zou and Banane, remained at the zoo as they were too aggressive to be integrated with the infants. The intention was to move them at a later date.

That evening we congratulated ourselves on a successful operation by visiting the local pizza restaurant. It also happened to be my birthday, so it was a double celebration.

The next morning we met at the zoo again. It was deserted, as most of the staff had decided against going to work and wisely stayed in their homes. But Jean and Ludovic, another Congolese chimp keeper, were both there as usual. Ludovic had agreed to accompany the chimps and stay and work at the sanctuary. Jean was going to remain in Brazzaville and look after the three adult chimps and all the other animals. We loaded the crates on to the back of a lorry, drove to the airport and boarded a plane. Accompanying me on this journey were Graziella, Cedric and his wife, Ruth, Ludovic and David Bowerman, a journalist who was filming the whole trip. Cedric, Graziella, Ludovic and I each held two of the smallest babies in our arms. I noticed that Ludovic seemed extremely tense and asked him if he was OK. He told me that this was his first time in an aeroplane. As the jet engines roared into action the babies clung tightly to us, while Ludovic clung even tighter to them. Then, one by one, the babies fell asleep.

The sign of blood returning to Ludovic's relieved face announced our successful landing at Pointe-Noire. As the plane taxied along the runway it was forced to jam on its brakes. I peered out of the window to see what the problem was. Dozens of Congolese citizens were crossing the runway, as one would a high street. As soon as they had passed, the plane continued and eventually reached its destination in one piece. Paul Actel, who was managing the Tchimpounga Sanctuary, had arranged the necessary clearance for the chimps. The lorry was on time and after loading the chimps on to the back, we set off on our lengthy journey to M'Pili.

The drive was long, bumpy and very hot. We all stood on the back of the open-top lorry beside the chimps. There was no shade, no clouds in the sky, and soon I was burnt to a crisp.

During the final hour of the journey we drove through a eucalyptus forest. These trees are indigenous to Australia, and it was disappointing to see them here in Africa. Even more distressing was the fact that great swathes of virgin rainforest were being cleared in order to plant even more of these exotic trees. Eucalyptus is popular because it grows extremely fast, straight and tall. It is perfect for use in the timber industry, because it grows so rapidly, and can be harvested within five to seven years. Other than hunting, one of the biggest threats to the survival of chimpanzees is deforestation and fragmentation of populations and their habitat. This threat can take the form of small-scale farming, such as slash-and-burn agriculture, which slowly erodes away the forest boundaries, or can be a much larger, devastating phenomenon, such as commercial logging.

Slash-and-burn is a crude farming practice in which forest is cut down and the area burnt and cleared. Crops are then planted for two or maybe three years. Because the topsoil is so thin and infertile, the farmers are forced to move on and clear new areas of forest, leaving behind them largely infertile land. The logging companies, most of them foreign, have a direct and indirect impact on habitat degradation. They clear large areas of forest and build roads, which allows settlement and in turn poaching. Their labour force, often many thousands strong, becomes a lucrative market for commercial hunters selling their meat, while the logging trucks provide a means of transporting the meat to other markets.

But the commercial hunters also have a negative impact on the local communities who depend on forest resources for their livelihood. It is estimated that commercial hunters take around ten times more bush meat than local subsistence hunters do. While the forests shrink and the cultivated areas expand, the chimp communities become fragmented and separated from neighbouring groups. Some of these communities become isolated and too small to be viable as long-term populations. These relic populations have little hope of ever coming into contact with other communities. In chimpanzee society, males always stay in their natal group, whereas females usually emigrate to neighbouring communities when they become sexually mature. This allows genetic and cultural transfer and prevents inbreeding.

After about two hours we arrived at M'Pili, tired, sunburnt and extremely thirsty. M'Pili is a small, remote village surrounded by the immense virgin Mayombe Forest. The local villagers were amazed to see all these chimps and so many *mondele* people — Lingala for 'foreigners'. Word soon got around and villagers came from far and wide in search of work or simply to watch. The sanctuary consisted of ten roomy cages, adjoining a large central play yard. This led to 20 acres (8 hectares) of rainforest, enclosed by a solar-powered electric fence. About 100 yards away was the sanctuary manager's house, which consisted of two rooms, a bathroom and kitchen.

Paul had organised a stash of local Primus and Ngok beers, and his Congolese cook had prepared a delicious meal of beef stew, manioc and plantain, so we partied into the night. Cedric and Ruth slept in the spare bed, and the rest of us slept on mattresses on the dining-room floor. The bright lights powered by a generator attracted all the insects imaginable (and many more unimaginable). These poured in under the door, into the room where I was sleeping. The room was an entomologist's paradise — only I was no entomologist and didn't want to be. I had no mosquito net, so I was there to be eaten. Some of the ants that crawled on to my mattress were spotted and over an inch and a half long. There were thousands of them. I sprayed my hands, arms and face with insect repellent and after double-checking that there was nothing in my bed, crept in. I awoke the next morning to find my feet and ankles covered in bites. I added feet and legs to my list of body parts to be sprayed the next evening and then had a cold, refreshing shower.

The vegetation inside the chimp sanctuary was dense and lush. Tall trees broke through the dense undergrowth, which at times seemed virtually impenetrable. I decided to cut some foot trails through the forest for the chimps. Before I set off, the cook warned me to watch out for snakes, especially green mambas. He told me that you can tell when they are about to strike because they prod you with their nose just before biting. I was always a bit sceptical of this kind of story, but he sounded convincing enough. So off I trekked.

After about an hour of hacking my way through the undergrowth, I came across an immense fig tree supported by impressive buttress roots. The tree was covered in lianas, and among their twisted woody stems was a mass of giant cobwebs. As I stood marvelling at the eerie beauty of these I felt a sudden sharp pain on my forehead, much more painful than any wasp or bee sting I had experienced. Something had bitten or stung me. It must be either a spider or, worse still, a snake, I thought. Wondering if the creature was still on my head, I thrashed around wildly, sending my machete spinning off into the air. It landed some distance away, scaring a couple of large fruit bats out of a thick mass of vegetation enveloping another tree. They flew around the trunk a couple of times, before disappearing into the dark forest. After failing to retrieve the machete, I headed back for the house. But, in my blind panic, I strayed from the trail that I had cut and was soon lost.

Realising that I was in the middle of nowhere, a long way from any doctor and perhaps mortally bitten on my head, I suddenly became acutely aware of my vulnerability. With the pain increasing by the second and with my heart rapidly beating the supposed poison deeper into my brain, I tore through the forest like a madman. Luckily, I soon stumbled back across my original path and found my way back to the house. I rushed into the bathroom to look in the mirror, expecting to see a grossly disfigured face before me. To my great relief the bite didn't look as bad as it hurt. There was just a small, red mark, and the pain seemed to be subsiding. Graziella found my plight somewhat amusing and told me that it was almost certainly a wasp or maybe a hornet. Feeling a wee bit foolish, I tried to explain to her that obviously I had presumed it was a wasp all along.

The next day I travelled to Pointe-Noire Zoo to deliver the chimpanzees to the sanctuary. There were nine in total, including several adults, and I was forced to anaesthetise four of these using a blowpipe and darts loaded with the drug ketamine. One of the adult females had just given birth a few days before in her tiny cage. This was probably the first chimp ever born in captivity in the Congo. Another of the adults was an ancient female called La Vieille (the Old One), who was slightly neurotic and extremely agoraphobic. In fact she wouldn't venture outside her small cage, even though the door was left open during the day and freedom was just one small step away.

A few days later I went on a trip into the Mayombe rainforest with Paul and his cook, who claimed he knew the area well. We were hoping to see some wild chimpanzees or gorillas. We spent all day there but saw nothing but giant land snails and insects. While swiping away at the thousands of gnats and mosquitoes buzzing around my face, I noticed that sitting on the back of my hand were what seemed to be two giant mosquitoes. I waved my hand about again, but they still didn't move. So I slapped them off with my other hand. Blood instantly poured down my hand and over my knuckles. Whatever these insects were, they must have pumped a powerful anticoagulant into my hand because the cuts bled as if I'd been slashed with a sharp razor blade.

Our trip into the forest proved a big disappointment. Other than a large branch crashing down just ahead of us, which we convinced ourselves was caused by a chimp running away, we saw precious little. I didn't even see any birds, even though I was surrounded by their melodic songs. They were too high up in the canopy.

Later that evening I was taking a shower when I discovered that I had two large sores on my lower back, each the size of a 50-pence piece and roughly where the waistband of my trousers would be. These turned out to be infected with *tumbu*-fly larvae. I could now feel truly at one with Mother Nature, having become a small part of the complicated life history of one of her many marvels. Paul had some spare Vaseline plasters, as he had recently had the same species of larvae on his arm. So I covered both areas and then went to bed, trying to put the maggots out of my mind. The sores seemed to take an eternity to heal fully and I felt sure their larval remains were still rotting beneath my skin.

I had heard so much about the beautiful beaches that adorn the Congolese coast that I wanted to witness them for myself before leaving. Paul was good enough to offer to drive me there. So we jumped into his Nissan Patrol and sped off towards the sea. From a distance the coastline looked magnificent — mile upon mile of white sand dotted with palm trees. But a closer inspection soon revealed an ugly secret. As I walked barefoot along the beach, side-stepping the numerous washed-up jellyfish that lay dying in the searing heat on the oil-stained sand, I imagined how this place must have looked not so long ago, when the sands glittered white and pristine. But the tarred beach, and the oilrigs on the distant horizon, soon brought my visions of paradise crashing down as I faced the reality of the twentieth century.

I returned to Brazzaville, and soon afterwards received a phone call from Jane. Paul was leaving, she told me, and she wanted me to take over the running of the sanctuary. But she needed someone in there quickly and wanted an immediate decision. I insisted on having a couple of weeks to talk it over with Audrey. Jane couldn't wait that long and sent Lucy, and later Cedric, over to the house where I was staying, to try to persuade me to take the job. I stuck to my guns and told Jane I would let her know as soon as I returned to England. She phoned me back the next day and told me about another sanctuary in Kenya, asking whether I would be interested in working there instead. I told her it sounded interesting and safer for my family.

7

Inhuman or Human Behaviour

I arrived back in England just in time to spend Christmas with Audrey and Oliver. I had missed my son a lot and was worried that, after two months away, I would find big changes in him and that he might have forgotten me. But I needn't have feared, for he was still as cute as ever. I rang the zoo to ask about Sophie and was told that she had suffered a few bites but was generally managing OK. I discussed Jane's offer of work in either the Congo or Kenya with Audrey. As she had never been to Africa, it all sounded very romantic and exciting. I'm sure that when we first met she could never have guessed where this crazy relationship was going to lead her. She was quite content to follow me to the ends of the earth. Our only real concerns were Oliver's health and education. So, bearing these in mind, we reached a decision, and that evening I rang Jane to inform her that we were pleased to accept the challenge and leave England for a life in the Kenyan bush. Jane told me that she would ask Geoff Creswell, who was originally lined up to work in Kenya, to take the position at M'Pili.

I then told her about Sophie's plight and that it would be a dream come true if she could somehow help me to get her out of her situation. Jane remembered Sophie from the time she had visited the zoo, and was very sympathetic. She promised me that, although there were no guarantees, she would try to speak to Dr Richard Leakey, the director of the Kenya Wildlife Service (KWS). Leakey is a renowned palaeoanthropologist, famous for his work on human evolution. The son of the equally famous palaeoanthropologists Louis and Mary Leakey (it was Louis who first sent Jane Goodall to Africa to study chimps), Richard had unearthed many important hominid fossil remains, including a *Homo habilis* skull in 1972 and a *Homo erectus* skull in 1975. Then, in 1984, his excavation team found the nearly complete skeleton of a *Homo erectus* boy, which was nicknamed 'the Turkana Boy'. The following year they had another major find, the first skull of *Australopithecus aethiopicus*. In 1989 Richard was appointed director of the KWS.

For Sophie there was now a first glimmer of light at the end of a long, dark tunnel. I returned to work at Chester Zoo at the end of December 1992, arriving early in the morning, and quickly went to see how she was. I was excited and a little apprehensive about seeing her again after such a long absence. At first I couldn't see her anywhere. I called out her name and there was no response. I called out again and a hessian sack suddenly shuffled along the floor from out of a dark corner. I called her name again, and the sack eventually shuffled towards me. Sophie was inside it, with the open end tucked firmly under her feet. She was completely hidden. It was pitiful and my heart sank. For a while she refused to come out of the sack, and sat there rocking back and forth. As her back banged against the wall, her broken whimpers made a distressing sound. I separated her from the chimps and eventually she emerged. A strong stench of pus filled the still air. I was shocked to see that she had again been severely attacked while I was away. She was badly bitten on her finger and hand, which was swollen to twice its normal size. I could see that her left toe was broken and was pointing vertically downward, preventing her from walking properly. Her back was also swollen and covered in bites and scratches. Her lips were cut and more of her teeth had been broken.

If that wasn't enough, she had to be anaesthetised the following day to remove the broken piece of her remaining front incisor. She had now lost five of her front teeth — two upper and three lower — with another of her few remaining teeth broken in half. The middle finger of her right hand seemed to be broken. The vet lanced the abscess on the back of this hand. She also had over a dozen bite marks on her back, along with many deep scratches almost a foot long. She was put on amoxycillin and given Calpol for the pain. Three weeks later Sophie was taken to the vet's surgery, where she was anaesthetised again. They took an X-ray of her wounds. Sadly, they were forced to amputate the upper half of her injured toe as it was causing her problems walking. The X-ray confirmed the fracture to her right-hand middle finger and also showed that her left hand was badly broken. In fact the break was so severe that the vet was able to put his finger between the two gaps in the bone.

We again kept Sophie in the kitchen. It was two days before she could walk properly. But gradually her confidence and character began to return, and she seemed happy to be away from the other chimps. So, after a few weeks, we decided to reintroduce Sophie, but initially keep her with just Meg.

Some days later I arrived at work early and was surprised to hear a strange noise. It sounded like a baby chimp crying, only this was a sound I had never heard before. I thought that Florin, who was heavily pregnant again, must have given birth. I looked towards the sound and saw a baby crawling on the floor with its head down. Florin must have abandoned her baby, I thought. With great difficulty the infant then raised its head. As it lifted its head feebly towards me, I realised, to my extreme horror, that there was no face. It had been completely bitten off. The head slumped back to the floor again and the baby continued groaning in pain. I could see, by its size, that this was not a newborn baby. Yet it was impossible to recognise this faceless chimp, which moaned constantly as it crawled about the floor.

I rushed around all the beds with my heart in my mouth, thinking it must be Sophie. My blood was pounding and my legs felt completely weak as I searched for her among the crowd of chimps. I called out her name but couldn't see her anywhere. There was Meg, but where was Sophie? 'Oh no, oh no,' I cried. The other chimps must have somehow got in with her, I thought, in my despairing panic.

But then, to my immense relief, I saw little Sophie sitting quietly and nervously in a dark corner. She had been with Meg all along, but I hadn't noticed her hiding in the darkness. She must have been terrified by all the screaming and confusion during that explosion of violence and madness. I ran around the building counting the remaining infants, to determine, by a process of elimination, who this tragic victim could be. All the infants were there except for one — Enya. I rushed over to check on Mandy and was disappointed to see her hanging on the bars, without her baby, blowing raspberries at me and asking for food. She seemed totally unperturbed by what was happening. Maybe she didn't recognise her own faceless baby or perhaps her maternal instincts were so weak that she just didn't care.

Unlike Mandy, two of the male chimps, Nicky and Boris, were both distressed and were sitting on either side of Enya, trying to pick her up. Despite their gentleness at this moment, it was quite likely that either of these two was the culprit. I was in a complete state of panic and shock but I managed to separate all the other chimps from Enya, except for the two males, who wouldn't leave her side. Eventually I had to fire water at them from a power hose to separate them.

My hands were shaking as I unlocked the door. I gently lifted Enya and she immediately clung on to me tightly, soaking me in her blood. It was like something out of a horror film. She had no left eye, her right eye was turned into its socket, there was no nose, no upper jaw, and her lower jaw was hanging down around her belly button by a thread of tissue. Her tongue, no longer supported by the jaws, flopped down her neck, and because she had no mouth or palate the cry she was giving off was a plaintive croaking sound. Despite her horrific injuries, her grip was very strong and it was tragic to see that she was very much alive. I was devastated.

I carried her in my arms and was driven to the vet's surgery — a journey of some 20 minutes. Throughout, she held tightly to my blood-soaked shirt, only relaxing her grip every few minutes as she passed out, then regaining consciousness a few moments later. I carried her into the surgery, where, mercifully, she was put out of her misery.

For several months Enya's mutilated face haunted me. At night, flashbacks would cause me to instinctively cover my eyes with my hands whenever her image entered my mind. No animal should ever have to go through such horrific injuries and suffering, and I wished that I'd never witnessed it.

Afterwards it was like a murder investigation. For weeks we suspected various chimps. Who was the killer? And would he or she strike again? What exactly happened and why Mandy seemed so indifferent will always remain a mystery. Could she herself have harmed her own baby? I doubt it, but then with chimps anything is possible. Even though I realised that it was unfair to apply human morality to another animal, I could never really feel the same affection for Mandy afterwards.

Infanticide represents the darker side of chimpanzee nature. As in human society, it is a relatively rare occurrence, and among chimps is usually committed by males, although females have also been known to kill and even eat babies. One possible explanation is that it is an abnormal response to overcrowding. Alternatively, it may be a male reproductive strategy, as the mother will come back into oestrus, or fertility, if her baby dies.

Chimpanzee society in the wild has developed a way to counter infanticide. Chimps live in an unusual fission-fusion social system, in which all members of the community are never seen together at the same time. This behaviour is believed to have evolved because the chimps need to both defend their territory and forage for the fruit that is distributed in patches throughout the forest. Because there isn't enough fruit for all members of the community to be found in one place at any one time, the chimps are forced to split, or fission, into subgroups, as the energy expended during travelling long distances in search of sufficient food for all the members would not make it cost effective in terms of expenditure of energy.

However, the subgroups need to regroup, or fuse, periodically in order to maintain their social relationships, which are important for effectively defending both their territory and other resources, such as food and reproductive females. Because of the nature of this fission-fusion society, chimpanzees from the same community may have no contact with each other for long periods. When they do eventually meet, they need to rapidly re-establish their previous relationships, in order to calm a potentially aggressive situation. They achieve this through formalised greetings, in which one animal presents its rear submissively to the other, who then either mounts the presenter or reaches through its legs to touch its genitals. This greeting is then normally followed by grooming.

Unlike males, who never leave their group permanently, females often migrate to neighbouring communities. If a female migrates with her infant, there is a chance that the infant will be killed. However, females share sexual favours with many males, any of which could be the father of her infant. Consequently, the only time that a male knows for certain that an infant is not his is when he meets a female that he has never seen before. If that female happens to be carrying an infant, then the male may well attack the female and kill the infant, with the result that the female comes back into oestrus.

Enya's death may have been caused by one of several factors. It may have been a result of having too many chimps in a confined area; or simply an aberrant action by an individual — either male or female; or it could, quite plausibly, have been due to the disruption of the group, caused by Sophie's introductions. Sadly, we will never know.

Shortly afterwards I was officially offered the job of setting up and managing the chimp sanctuary at the Sweetwaters Game Reserve Ranch in Kenya. The managing director of Lonrho Agri-business, Russell Clarke, flew to England and visited me at Chester. Russell was a qualified accountant, but had previously been an honorary warden in Zambia and was passionate about wildlife. I took him on a tour of the zoo and showed him Sophie. He had heard about Sophie from Jane, and agreed it would be great if she could also accompany me to Kenya and said he would do what he could to make this possible.

In a peculiar and ironic way, Enya's tragic death was one of the main turning points in Sophie's life. If the chimps could do this to Enya, who was born in the group and was being reared by her mother, then they could easily do this and more to an orphaned 'outsider' like Sophie. All along I had been praying for some justification for approaching the zoo management with a genuine solution and alternative to Sophie's plight — an opportunity to end the mental torture she was enduring.

Nick Ellerton agreed to let Sophie follow me to Africa. It wasn't the zoo's policy to sell animals, especially endangered species such as chimpanzees. Chimpanzees are a CITES Appendix 1 listed species. CITES is the Convention on International Trade in Endangered Species, and all animals are categorised according to their status in the wild. Chimpanzees, along with species such as gorillas, elephants, black rhinos and giant pandas, are Appendix 1 listed. So the zoo decided to let her go to Kenya on 'permanent breeding loan'. It was a wonderful gesture on Nick's part and I will always be indebted to him. Jane also kept her promise and helped by persuading Richard Leakey to give his approval and allow Sophie into Kenya.

I couldn't believe it. After all the suffering, it seemed fate had brought us back together, and it felt like a miracle. But there was still a long way to go, and because of all the complicated CITES paperwork, there was no way of knowing how long it would be before Sophie would be allowed on a plane to Africa.

That night I lay in bed smiling from ear to ear as I thought about uniting my family again in a far-away land. It had always seemed an impossible dream, but maybe now my dream was about to come true. Still smiling, I closed my eyes and dreamt of Africa.

8

Karibuni Kenya

We decided that I should go to Kenya initially on my own and that Audrey and Oliver would follow me two months later. This would allow Audrey to sort out all the business formalities, including renting out our house. She also had to make arrangements for our three cats, as they too were members of the Smith household and would soon be following us to Africa. But first they needed to spend a month's quarantine in England. We also had to sell or find homes for all our other animals — something which couldn't just be done overnight.

Sophie's integration was put on hold and she remained on her own at Chester for several months while her CITES paperwork was sorted out, which meant obtaining such things as export and import papers. It was also necessary to get her vaccinated, have a wooden crate especially made, and arrange her quarantine and transport to Kenya. It was a lonely period for Sophie, and from time to time Meg was kept in with her for company.

My plane landed at Nairobi's Jomo Kenyatta airport on the last day of May 1993. I was met by Simon Barkas, the general manager of Ol Pejeta Ranch, where Sweetwaters Game Reserve is situated. Simon was a second-generation white Kenyan farmer, a couple of years younger than me. As we drove away from the airport and through the industrial area, we passed under a large sign across the road with the words 'Karibuni Kenya' (Swahili for 'Welcome to Kenya') emblazoned across it. As we continued along this road towards the impressive line of skyscrapers that marked the centre of Nairobi, I wondered where my own life was now heading.

Once we had left the city centre we headed north on the three-hour journey towards the town of Nanyuki in the District of Laikipia. Laikipia is an enormous level plateau stretching from the slopes of Mount Kenya and the Aberdare Mountains to the edge of the Great Rift Valley. Essentially a wilderness the size of Belgium, it supports large numbers of animals, including one of the largest populations of elephants, and is the last stronghold in Kenya of the endangered black rhino. Our drive gave me a chance to take in some of the country's magnificent scenery. The climate, vegetation and landscape changed dramatically throughout the journey. As we continued further north, so the climate gradually became less humid. Then, as we reached the halfway point, near the town of Nyeri, the fertile red clay and the lush greenery dramatically changed to black 'cotton soil' and a drier more sparsely vegetated landscape. So named because cotton grows well in it, this soil absorbs huge amounts of water, becomes very slippery and sticky and is famous for making roads impassable in the rainy season.

A large signpost just outside Nanyuki marked the equator. A few huts were selling masks and souvenirs to eager tourists who had stopped by the sign for a photo before continuing through the town and turning left at the crossroads towards Nyahururu. After a few hundred yards the tarmac abruptly disappeared, and for the remaining 12 or so miles of our journey we followed a dusty, potholed dirt track before finally reaching the entrance to the ranch.

We then drove the short distance to Simon's house in time for lunch, and there I was introduced to his wife, Lulu, and baby daughter Daniella. Afterwards Simon gave me a tour of Ol Pejeta, proudly showing me the ranch's impressive herd of pedigree Boran cattle. These medium-sized beef cattle, grey, red or fawn in colour, were originally developed from the cattle of the Borana people of southern Ethiopia. They are well adapted to living in harsh conditions, as they are resistant to heat, ticks and eye diseases and can survive with little water and on poor-quality feed. The females have a reputation for being highly fertile and excellent mothers.

Ol Pejeta Ranch is owned by the multinational company Lonrho. Founded by the late Tiny Rowland, Lonrho is a large conglomerate of hotels, agricultural and mining companies spread throughout Africa. Ol Pejeta was formerly the private holiday home of the Saudi Arabian international financier Adnan Khashoggi. However, he allegedly ran up huge gambling debts at Lonrho's casinos, and when he was declared bankrupt, lost his property to Lonrho. Khashoggi's former residence, the Lodge, is rented out by Lonrho to wealthy tourists and includes much of the original furniture, such as two massive beds, 13 feet wide, and a dining table above which hangs a bamboo pirogue that can be lowered by a pulley. Usually this canoe would be full of fruit, but legend has it that Khashoggi would entertain his dinner guests by lowering it down to the table with a naked girl hidden inside.

For the first week I stayed at Simon's home high up on the ranch. The ranch was immense, 110,000 acres (44,500 hectares), bigger even than Kenya's extensive Samburu National Park. It was coming to the end of the long rainy season, so everywhere the vegetation was lush and the animals fat and healthy. Within the ranch lay the 25,000-acre (10,100 hectares) game reserve, home to 'the big five': lion, leopard, buffalo, elephant and rhino. It was also inhabited by a host of other species, both animal and bird. But pride of place went to the 20 black rhinos. There were as many species of wild animal on the ranch as in the reserve, but the rhinos were only found in the reserve. Sweetwaters attracted about a thousand tourists every month. Many would stay a few nights at the luxury Sweetwaters Tented Camp. Although it lay inside the reserve, the Tented Camp was run by Lonrho Hotels, and was managed separately from the ranch and the reserve. Although the accommodation was technically tents, these were quite lavish inside, with en suite bathrooms. All the tents overlooked a waterhole that was lit up in the evening, allowing the guests to view nocturnal game from the comfort of their own quarters.

After a few days I was handed the keys to a Toyota Land Cruiser which used to belong to Khashoggi's ranch manager. I was also offered the choice of two houses in the reserve, at either Pelican Dam or Spoonbill Dam. Both dams were private and off limits to tourists. The house at Pelican Dam was the larger of the two, but was less secluded as it was close to the sanctuary's perimeter fence. So I chose the smaller but cosier house at Spoonbill Dam, which was hidden among the thick bush in the middle of the game reserve.

Spoonbill was a beautiful, African-style house, circular in shape and built from a mixture of mud and stone, with a roof of thick thatched grass. Inside there were two bedrooms, two bathrooms with toilets, a study and a very large dining room. The kitchen was built outside the house so that the thatch didn't catch fire. A small Lister diesel generator supplied ample power to the house. Outside was a two-acre (0.8 hectare) garden with a fenced-off vegetable plot. To the front of the house were magnificent views of Mount Kenya, and to our right the Aberdare Mountains. At the rear was what was probably the property's most attractive feature: a magnificent dam, one of six in the reserve, measuring just over an acre (0.4 hectare) and for the use of just ourselves and the wild animals. The back door of the house led on to a large veranda overlooking the dam.

Our dam, which got its name from the abundance of spoonbills, pelicans and other water birds that frequented it throughout the year, was of tremendous importance to the many birds and animals in the sanctuary, especially during the dry season. The dam had a healthy population of fish and crustaceans, mainly barbel, crayfish and freshwater crabs. These were a further attraction for the birds, which flocked there in large numbers. It was the wading birds that made the place special. I had a casual interest in ornithology, and so spent many hours familiarising myself with the many species in my garden.

During my first month alone, the dam was visited by a variety of bird species, including fish eagles, ospreys, pelicans, crowned cranes, cattle egrets, white-necked cormorants, pied kingfishers, sacred ibis, hadada ibis, marabou storks, saddlebill storks, spurwing geese, red-billed ducks and, of course, spoonbills. Spurwing geese are large birds which, from a distance, resemble black swans. The two that were nesting on the water were the only pair in the reserve. Animals would come from every direction to drink or to bathe in the water. This meant rich pickings for the lions which lay in wait to catch their prey off guard. It was a wonderful place for our child to grow up and a true education in itself.

The climate in Laikipia was perfect: hot but not humid during the day and cool in the evening. In fact the evenings could get quite chilly and on the occasional morning you would wake up to a frost on the lawn. I would while away the evening hours warming myself in front of a log fire with a local beer in my hand. The biggest luxury of all was that, owing to the altitude, there was no malaria, so I didn't need to take any prophylactic medicine.

One of the reasons why I had chosen Kenya rather than the Congo was because I felt that it would be a safer environment to bring up my family. However, I soon found out that Kenya was no Garden of Eden. Many of the local hotels in Laikipia were being systematically robbed by an armed gang of bandits. The raiders seemed to be hitting each of the major hotels one by one, and the police seemed powerless to stop them. A few days after I arrived they attacked Mount Kenya Safari Club, in Nanyuki, also owned by Lonrho. We were now all on red alert, fearing that the Tented Camp would be the next target. I began to wonder what I was bringing my family to.

However, during the attack on Mount Kenya Safari Club, one of the bandits was wounded by an arrow fired by the Masai security staff. He escaped but later made the mistake of going to the local hospital for treatment. One of the hotel's security men, who had been injured in the attack and was also being treated there, recognised him and raised the alarm. Now there was a lead to follow. But soon afterwards it was discovered that the gang consisted mainly of members of the local police force. Consequently, the entire staff were moved to another district. They weren't arrested or sacked — just moved somewhere else! The attacks in our area ceased overnight. Throughout this unsettling period I didn't let on to Audrey about all these problems as I didn't want to worry her before she had even set foot in the country.

At Spoonbill, I didn't need much reminding that I was living in the middle of the African bush. Each night I would religiously check beneath my bed and under the sheets for any snakes or spiders. Then, after carefully placing my shoes on a chair with my socks tucked inside, I would slip into bed and lie there listening to all the sounds from outside my window: zebras thundering past barking their 'kwa-ha, kwa-ha' alarm calls, buffalo splashing in the dam and elephants emitting their low rumbles. Most nights I would hear hyenas prowling around producing their eerie contact calls, just part of their impressive repertoire of vocalisations. One evening I heard them laughing, their cries intermingled with loud roars from a lion. They must have been trying to steal the lion's kill. All this was taking place just a few yards from my window. It was like listening to your favourite music before going to sleep. However, although I heard hyenas yelping most nights, this turned out to be the only time I ever heard their strange laugh.

On one of my first nights I was woken by something biting my left arm. Initially I thought I was just imagining things and remained motionless for a few moments. Nope! There was definitely something on my arm. Keeping my left arm completely still, I fumbled in the dark with my right hand for my torch, and shone the light under the blankets and on to my arm. Crawling through the hairs of my arm were five large, black ants, and two of them were biting into my skin. I pulled back the sheets and shone the light over the rest of my body, only to discover that I was sharing my bed with an army of safari ants, also known as driver ants. They formed a line of thousands that went across my pillow, past my shoulder, down my arm, along the side of my left leg, out of the bed by my feet and then down the leg of the bed to the floor.

My moving and shining the light immediately disturbed them, causing them to disperse in a mad rush everywhere in the bed. Oops, time to go! I jumped up in a flash and scanned the floor with my torch. Every couple of inches of it seemed to be occupied by at least one ant. I hopped around the house with my torch like a madman barefoot on hot charcoal. They were in every room except one. Luckily, it was the spare bedroom.

When safari ants go on the march, nothing steps in their way, and you are defeated by their sheer numbers. These nomadic animals are one of Africa's most feared predators. They move in columns numbering several million, in densely packed lines along the ground, in search of prey such as millipedes, spiders, beetles and anything else that gets in their way, including reptiles and small mammals. They have been known to kill chickens and even small babies. The columns of ants are protected by large 'soldiers', equipped with powerful jaws, and these stand to one side with their jaws held menacingly open.

Then I found the ants were beginning to invade the spare bedroom, although so far there were only a few dozen or so at the most. I brushed these out quickly and collected some ash from the smouldering remains of the fire. Then I laid a thick wall of ash, one inch deep, across the bottom of the door. This seemed to do the trick, so I climbed into bed. I lay there feeling itchy and wondering if my wall of ash would hold out.

The next morning I was woken by the melodic song, from outside the window, of the resident golden oriole. All the ants had gone and there was no sign of their visit, other than that the floor was spotless. It seemed that these house-proud insects had gathered up every last spot of dust, doing my housework for me as I slept. As I swept up the ash from my door, I couldn't help thinking that on this night I had won the battle, but maybe not the war, as I was sure they would be back. But that was what living in the African bush was all about: it wasn't just about the large mammals and birds that people normally associate with Kenya; it was also about the spiders, the many snakes, the biting insects, the mud, the dust and those damned ants and ticks.

If I thought my house had a lot of bugs, then outside there were many more. I once walked a few hundred yards through the savannah, where tall red oat grass dominated, and my jeans were soon covered in grass seeds. At least that's what I thought they were. As I stopped to brush these tiny black specks off my trousers. I quickly realised that they were all moving rapidly in the same direction — up towards my groin. I was in fact covered in hundreds of 'pepper ticks', the term used to describe newly hatched ticks, which are so small that they look like grains of pepper. They were all very hungry and looking for their first meal.

Sweetwaters was infested with the brown ear tick. The adults, as wide as the blunt end of a pencil, are capable of remaining dormant without feeding for more than 14 months. They wait patiently on vegetation for an unsuspecting animal or human. Then they explode into life, grabbing hold of their prospective host as it walks by. They prefer to find a warm region of your body, such as your armpit or groin, in which to nestle and burrow their head beneath your skin. Before they begin feeding, they regurgitate their saliva into your blood. The saliva contains an anticoagulant and so prevents the blood in the area of the wound from clotting. There they remain, engorging themselves on your free-flowing blood. You tend not to feel them while they are feeding, as along with the anticoagulant their saliva contains an anaesthetic, which numbs the pain. It is after they have gone that the wound begins to itch like a mosquito bite, partly because of the body's reaction to their saliva. It is always a mistake to pull them off, as this causes them to leave their head beneath your skin, and the area can easily become infected. Then they itch like crazy.

Ticks are a big concern in Africa, as they transmit many diseases, such as East Coast fever in cattle, Nairobi tick disease in dogs and rickettsial tick fever (or typhus) in humans. Each evening I would inspect every part of my anatomy for ticks before going to bed. It was a rare occasion when I didn't find at least one small pepper tick somewhere on my body.

* * *

During my first month I set about engaging some house staff. The ranch employed several hundred Kenyans from many different tribes. First I took on a Kikuyu *askari* (Swahili for night guard) called Steven Waweru. Then I employed a tall Turkana man, Daniel Auwalan, as a gardener. Daniel was already working in the reserve and had a reputation for being strong and brave.

The staff houses were situated at the rear of our compound. They consisted of simple round huts made from a mixture of mud and cow dung, with a thatched roof of reeds cut from the marsh in the reserve. There was typically a small window, which allowed a narrow beam of light into the gloomy interior. Inside, the huts were sectioned off into two rooms, the larger being the all-purpose room and the smaller the bedroom. Old newspapers would be pinned to the insides of the mud walls as wallpaper or decoration. *Jikos*, small charcoal cooking stoves, would be kept constantly alight, making the rooms pleasantly warm but unpleasantly smoky.

Many of the staff's social activities took place while they sat outside the front door of their homes. Most evenings they would gather with friends and sit and chat and share a drink or a meal. A typical dish was *sukuma wiki*, meaning 'stretch the week'. This is a stew made from a mixture of spinach or cabbage, tomatoes, green peppers, onions and any left-over meat, and would be eaten with either *ugali* (maize porridge) or chapatis. For special occasions they might prepare *nyama choma*, or grilled meat. They would go to bed early, at around eight in the evening, then get up at the crack of dawn.

Pay day, at the end of the month, was a time of much drinking and merriment on the ranch. My house staff would invariably end up drunk that evening. But as long as it didn't affect their work the following day, I didn't mind them unwinding. Besides, there was little I could have done to stop them. They were all hard-working and I would always find them up before me in the mornings. Steven would be preparing a fire in the outside wood burner to provide hot water for the house, and Daniel would be busily slashing the grass low around the house to keep down the number of ticks and, more importantly, so that you could see any snakes hiding in the grass.

Sweetwaters had a healthy population of snakes. There are more than 420 species of snake in Africa, of which only about 90 are poisonous. Fortunately, most of the species in the reserve were non-poisonous, although among the more common was the highly poisonous puff adder. These thick-bodied snakes grow up to four feet long. On the basis of the number of their victims, they are generally regarded as Africa's most dangerous snake. Lethargic and slow-moving, they are easy to tread on by mistake. At Sweetwaters they caused a major problem to cattle on the ranch.

Most of the local Kenyan staff greatly feared all snakes, believing that the only good snake was a dead one. But if snakes can't live unharmed in a game reserve, then where on earth can they live? I used to see snakes regularly and kept a running total of the ones I encountered on my travels within Sweetwaters. I seemed to develop an eye for spotting them and counted 94, of which six were found in my house. I tried explaining to my staff that snakes are useful to humans because they help control the rodent populations, and that they shouldn't kill them but catch them and release them in the reserve, far away from their houses. Unconvinced, they nodded in agreement, and I could see they thought I was quite mad.

On reflection, trying to get my staff not to kill snakes was a bit like asking them to become vegetarians on the basis that it was cruel to eat meat. I knew I had failed when a few days later I caught Daniel carrying a freshly killed egg-eating snake. He had killed it with a single blow to the head and was about to throw it away into the bush. The snake still had an unbroken egg inside its mouth. I never did fathom out how he had managed to kill it without breaking the egg. I told him disappointedly that this snake was harmless and that when I had said to catch them and carry them to the bush, I had meant that they should be alive. He then pointed to the egg inside its mouth as if to say, 'Try telling that to the egg — eggs have feelings too.' I could see I was fighting a losing battle.

One of my first tasks at Sweetwaters was to choose a suitable spot for the chimp sanctuary. I spent much of my time driving through the reserve looking at potential sites. In the end I chose an area by the Ewaso N'yiro River. This perennial river, with its Samburu name meaning 'brown water', is one of the largest rivers in Kenya. Flowing from south to north, it begins its journey in the Aberdares and then flows 125 miles over the Laikipia Plateau, through Samburu National Park, before vanishing into the Lorian Swamp in the east. The Ewaso N'yiro is the main water catchment area for the whole Laikipia District.

Now that I had chosen the site, I set about designing a plan for the sanctuary. This consisted of a large building with a forested area of one square kilometre, surrounded by a 10-foot solar-powered electric fence. The fence would produce a mild shock in the form of a one-second pulse, similar to a cattle fence. It wasn't harmful in any way, only unpleasant, as I later found out to my cost. The Ewaso N'yiro conveniently bisected the sanctuary, with one side of 110 acres (44.5 hectares) for the adults and the other of 104 acres (42 hectares) for the infants. The plans were accepted by Russell Clarke, so I set about finding a building contractor. Simon recommended a local man who had previously done a good job for the ranch. That sorted out, I started thinking about recruiting staff for the sanctuary. I needed chimp keepers, maintenance staff for the electric fencing and night guards. I put the word out that I was looking for staff and it wasn't long before I was inundated with applications. After completing the interviews I settled on my team: Joseph Maiyo as Assistant Manager; Richard Mutai and John Kiama as adult chimp keepers; Dickson Kariuki, David Mundia and John Sungura as infant chimp keepers; Paul Wanyai in charge of maintenance; Timothy Mwangi and Antony Kamau as fencers; Charles Musasia as head gardener; and Samuel Kariuki, John Kabira and Vincent Gatete as *askaris*.

Soon after I arrived in Kenya I learned that Richard Leakey had been seriously injured in a plane crash after the single-engined light aircraft he was flying suffered engine failure. His legs had been badly crushed and, despite painful attempts to save them, both needed to be amputated below the knee. This also proved to be a significant blow to Sophie's chances of coming to Africa. Now, most of the decision-making concerning her was put on hold until Richard's return and there was no way of knowing how long this would be.

Despite this setback, the building of the sanctuary continued at a rapid pace. A team of 20 or so workers arrived each day. They initially set about clearing the bush for the fencing. This meant clearing a line of bush, two and a half miles long, down to its roots and then digging holes every ten yards for the fencing posts. Then a different team began working on the foundations of the chimp building.

Overseeing the construction work kept me busy, so I didn't have time to feel lonely when I was on my own. Once they began laying the walls, the building started taking shape and looking like my original designs on paper. We had bullet-proof glass specially made for the windows, and the ranch's workshop engineer, Ian Agagliate, did a professional job welding and fixing all the bars and slides into position.

Generally, there were few hitches, except for minor problems such as ensuring the door frames were symmetrical. I doubt if my presence at the building site was much appreciated by the foreman, as I was forever complaining that things weren't straight and making him redo the work. One morning I arrived to find they had fitted the frame of the main entrance door completely askew. It made the building look like a crooked house in a fairground. To make matters worse, the foreman insisted that it was straight. It was only after I made him measure both sides of the frame that he conceded that the left side was almost two inches higher than the right.

On 19 July I drove down to Nairobi to meet Audrey and Oliver at the airport. I was feeling very excited at being united with my family again and couldn't wait to drive them through the sanctuary for the first time. They had taken the night flight and were arriving early in the morning. It was a big move for Audrey. She was very close to her family, especially her mother. But she was excited, if not also a little apprehensive, about following her husband to a new continent on yet another adventure. Although born in Mauritius, she had never returned there or visited mainland Africa. Oliver, who was only two years old, had been very good on the plane, sleeping most of the way. Audrey hadn't managed to sleep at all. We booked in at the 680 Hotel and the first thing she did was to catch a couple of hours' kip. Then, in the afternoon, I took them for a stroll around Nairobi.

The following morning we checked out of our hotel and began the long drive to Nanyuki. We entered the game reserve and drove the 20 minutes to our house. Oliver's face was a picture — glued to the car window as we drove through a mosaic of differing habitats of acacia woodland and savannah. We passed numerous herds of animals, such as Thomson's and Grant's gazelles, impala, waterbuck, warthog, ostrich, eland, kongoni, oryx, jackal and common zebra. In amongst the bush, elegant Masai giraffe were feeding off the acacia tops, and in the distance a herd of elephants hurried towards one of the dams — and all this within less than half an hour of entering the reserve. It was the middle of the dry season and the parched grass offered little nourishment to the grazing animals. Many tended to retreat into the bush to find the last remaining tussocks of greenery hidden beneath the shade of a bush or to browse on the tiny acacia leaves. When we reached the private drive to our house, I watched Audrey's face light up as our new home came into view.

'Wow!' she exclaimed. 'It's wonderful.'

Steven and Daniel were waiting to greet my family. After we had unloaded the car and looked over all the rooms of the house, we strolled around the garden and ended up relaxing by the dam, where we uncorked a bottle of red wine.

Audrey soon settled into her new life in Africa and began adding many feminine touches to the house. We decided to employ a house girl, and after several interviews we selected a young Turkana called Grace. Audrey had brought with her some vegetable seeds from England and soon set to work arranging our vegetable garden. With its fertile soil, it wasn't long before it had become very productive. We were soon eating sweet potato, courgettes, aubergines, red and green peppers, red chillies, broccoli, spinach, artichokes, pumpkins, marrow, and, best of all, parsnips for our Sunday roast. We also had fruit trees, including bananas, papaya, tree tomatoes, strawberries and three different types of passion fruit.

A few weeks later our container arrived with a selection of our furniture, personal belongings, all my books and Oliver's toys. Oliver kept his toy chest safely under his bed. One day, after pulling it out, he rushed over to me in the study, yelling, 'Snake! S-s-s-snake!' I went over to see, and, sure enough, there, wriggling furiously in a panic under our bed, was a small blind snake, about eight inches long. We carefully collected it in a jam jar and released it safely outside. Fortunately, these small snakes are completely harmless. They have only rudimentary eyes and lack teeth in their lower jaw. They look very much like large garden worms, and I would even go as far as to say they are cute. They feed on ants and termites, so in some ways they were welcome visitors, although maybe less so under our beds than in the garden. Steven had probably carried it into the house among the firewood. He would collect any old bits of wood he could find lying on the ground, usually broken off by elephants. Apart from two more blind snakes, other animals he inadvertently carried into the house were a brown house snake, a black snake and a bushbaby. Hmm, maybe he didn't like us after all!

Every day seemed to be hot, dry, windy and very dusty. Driving through the reserve, I would leave a trail of dust in my wake about 100 yards long, like a scene from an old Western. I remember my friend Steve Hogarth, from the zoo, who had previously visited Kenya on holiday, warning me that, because of the wind and dust, the hair on my head would soon disappear and be replaced by thick nasal hair. It wasn't a very appealing thought and he wasn't far wrong.

We were now in the middle of the dry season and it was proving long and hard for the animals. July and August are also the windy season in Laikipia. The last bits of moisture in the grass soon disappeared under the combined efforts of the wind and scorching sun, leaving a parched landscape. For the animals it was now a battle of survival where only the fittest would live to see the rains. Two of the dams in the reserve had completely dried up. Each day and night animals came to our dam in their hundreds to drink. Large herds of zebra and eland congregated, and occasionally giraffe, waterbuck and rhino, and in the evening buffalo would arrive in herds up to 80 strong. These were easy times for the lions that lay in wait by the dam, hidden under the shade of the bushes, for their thirsty victims. Lions are in fact one of the most inactive animals, spending around 20 hours a day resting, avoiding the midday heat and conserving their energy.

After their quarantine period in England, our three cats finally landed safely in Kenya in August. They each came in their own very smart carrying boxes. I sorted out the formalities with the customs officials and then drove off with them meowing in the back of the car.

We kept the cats in the house for the first few days and they quickly settled into life in the bush, where they became excellent rat catchers. They took a little while to adapt to the daytime heat, and we needed to give them rabies injections, and flea collars to fend off the ticks. Each night we had to lock the cats in the house, as there were many much larger predators prowling around the garden after dusk. Now we were almost complete as a family, with just one member missing — Sophie.

We were already well into October and the Kenya Wildlife Service was dragging its feet over the Sophie issue. In Richard Leakey's absence, they almost reneged on the original agreement to allow her to come to Africa. I had to more or less begin again from scratch, justifying why I wanted her to come to Kenya and explaining that if she didn't come, then her survival chances would be minimal. No one seemed prepared to stick their neck out and make a decision. In the end I virtually had to beg them to honour Leakey's original verbal agreement. The KWS, however, were more concerned that by letting one non-indigenous animal into the country, they might be setting a precedent, which would allow people to bring in other exotic species. But, considering they had agreed to bring in 20 chimpanzees from the Jane Goodall Institute in Burundi, I found their argument somewhat flawed, not to say unsympathetic. Finally, Dr Jim Else, the deputy director, gave in to my pleading, saying that he was doing this more for my own peace of mind than for Sophie's welfare.

Although Kenya has an immense wealth of wildlife, there are no species of ape living in the wild, unlike in neighbouring Uganda, which has not only savannah animals but also chimpanzees and gorillas. There are two species of chimpanzee in Africa: the common chimpanzee, *Pan troglodytes*, which ranges in 21 countries across equatorial Africa, from Senegal in the west to Tanzania in the east; and the bonobo, or pygmy chimpanzee, *Pan paniscus*, which is endemic to the Democratic Republic of Congo and found only in the central Congo basin. It is estimated that the total population of common chimpanzees in the wild, of which there are currently four recognised subspecies, is in the region of 110,000, and of bonobos between 10,000 and

20,000. The different subspecies vary slightly in colour and size, and also in certain social behaviours.

Although all chimpanzees are CITES Appendix 1 listed, little is ever done by the local governments to enforce these laws. One of the problems facing governments is what to do with the chimps if they do confiscate them, as they have neither the money nor the facilities to adequately keep them. So most governments tend to turn a blind eye to the issue of illegally detained animals, and fail to implement their CITES laws.

So why was Kenya chosen as a suitable site to build a chimpanzee sanctuary, particularly as chimpanzees are not indigenous to that country? Although, in the wild, chimps range as far east as neighbouring Tanzania and Uganda, there is no evidence that they previously reached Kenya, presumably because of the impenetrable barrier created by the Great Rift Valley. This despite the fact that the Kakamega Forest, the only natural rainforest in Kenya, has seven species of primates, and is probably a suitable habitat for chimpanzees. Any sanctuary must be prepared to care for each orphan for its entire lifetime. Because chimpanzees can live for 50 years or more, this is a massive burden and an expensive commitment. However, there are three key reasons why Kenya is a more attractive proposition for a chimp sanctuary than most countries that have indigenous wild populations: it has tourism, which is needed to finance the running of a sanctuary; it has the infrastructure

Sophie and me,
Christmas 1990

Sophie's box

Meeting Oliver for the first time

Hello there!

Posing for the camera

What is this green stuff?

The family down by our wood in Shropshire

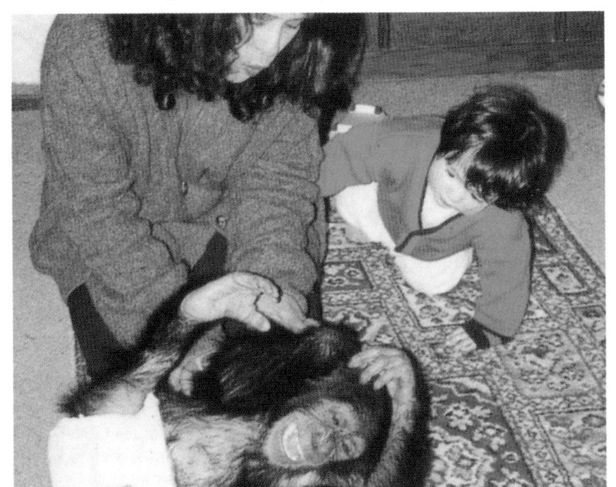
Being tickled by Audrey

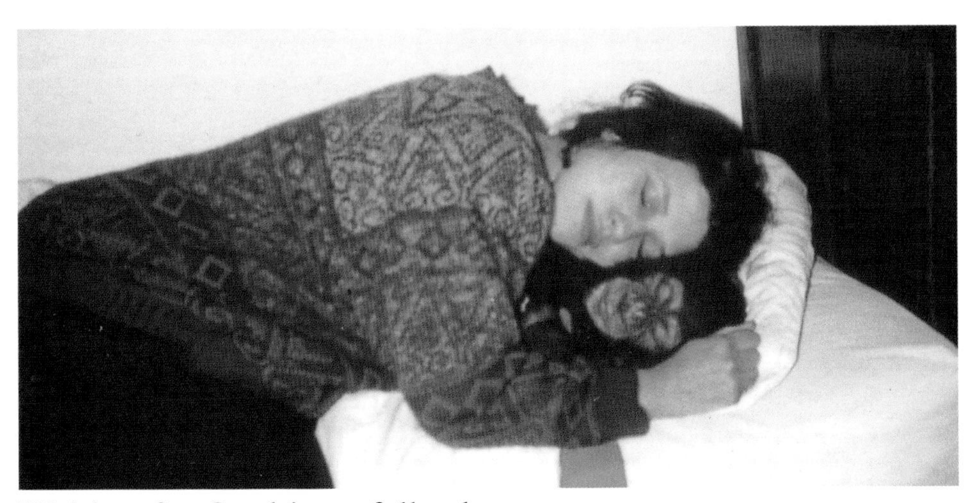
Waiting for Sophie to fall asleep

Taking Sophie to meet her real parents, Boris (far right) and Mandy (centre) – note the family resemblance between father and daughter

Oliver's and Audrey's emotional reunions with Sophie after she'd been back at Chester Zoo for six months

On our way to the sanctuary

Part of the group … Sophie and Tess are on the right

Sophie aged eight

Sophie meets Jane Goodall again, at Spoonbill

Cuddling Sabby the cat

Family portrait

Sophie improvised her own use of tools, slicing bark from a stick (left) or cutting wood with a machete (right) without ever being taught how

A midday sugar-cane snack

Enjoying her fifth-birthday present

Our house at Spoonbill Dam in Sweetwaters Game Reserve, Kenya

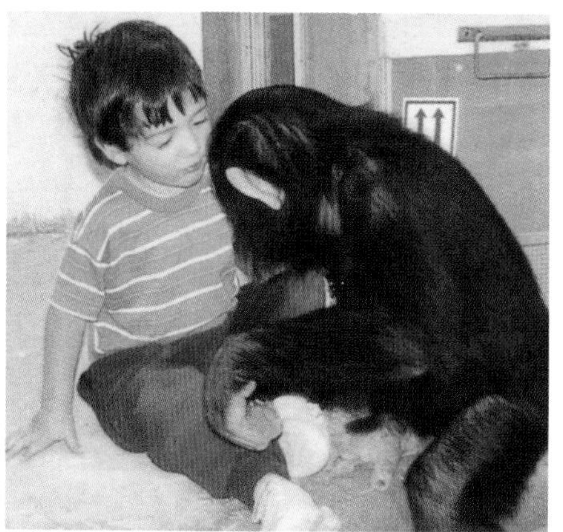

Sophie and Oliver reunited upon her arrival (in a wooden crate) at Sweetwaters

A tender moment between the two

Me with my two primates

to handle the tourists; and, importantly, it has, for Africa, relative political stability. Also, chimps are susceptible to most human diseases and ideally chimps exposed to humans shouldn't come into contact with wild populations. As chimps are not indigenous to Kenya, there is no danger of this happening.

⋆ ⋆ ⋆

Meanwhile Sophie was still living out her life at the zoo in England. It must have been a lonely existence for her, but at least she was safely away from the other chimps — though no doubt she believed that it was only a matter of time before the introductions would all begin again. I kept in constant touch with the zoo by telephone. Things had progressed a little but we had now reached an impasse. To obtain a CITES export licence for Sophie, Chester Zoo needed a CITES import licence from Kenya. However, the Kenya Wildlife Service were saying that they needed a CITES export licence before they could issue an import licence. We were stuck in a chicken-and-egg situation, and it was all very frustrating and rather pathetic.

Fortunately, Richard Leakey returned to the Kenya Wildlife Service after five months away, during which time he underwent major surgery. Sophie was quickly given her CITES export and import papers which would allow her to come to Kenya on a breeding loan. Not long after his operation Richard visited the chimp sanctuary. The roof was being built at the time, and a wooden ramp led from the ground steeply to the top of the building, allowing the workers to run up with their wheelbarrows full of wet concrete. Richard insisted on walking up this ramp unaided to view the sanctuary from the top of the chimp building. He had only just recovered from having both his legs amputated and was still getting used to walking using his artificial limbs. It was a struggle but he made it to the top unassisted. I could only admire his determination and will power.

Our first Christmas together as a family in Africa was a memorable one. In Kenya you don't get the commercial overkill that you do in Europe. But it is a special time of year for all children, so we headed off to Nairobi to buy our presents. Oliver was just over two and a half, and old enough to get excited about Christmas. As we were living in the bush, we followed the example of the other expatriates living at Ol Pejeta, and used a small whistling thorn acacia tree for our Christmas tree.

The whistling thorn tree has an interesting biology. It is an acacia that can grow to about 18 feet tall, but in most cases reaches around ten feet. It has extremely sharp thorns, about three inches long, which are paired, and at the base of each thorn is a hollow swelling, known as a gall, about an inch in diameter. When these galls become old the wind blowing through them causes them to make a whistling sound, hence the name. They normally become home to one of four species of ants, *Crematogaster mimosae, C. sjostedti, C. nigriceps* and *Tetraponera penzigi*, which make a small entrance hole in the gall and form their nest inside. These species are constantly at war for possession of trees and do not tolerate the presence of the other species on their tree. In fact *C. nigriceps* is so intolerant of other ants that it vigorously prunes off the tips of the branches and flowers to ensure that its tree doesn't grow into contact with a neighbouring tree. The ants obtain shelter and food from a sweet sap solution excreted from glands within the leaves called nectaries. The tree receives protection from the ants, which attack animals browsing on their host tree.

Although we were keen on having a Christmas tree, we weren't so enthusiastic about sharing our dining room with a colony of biting ants. But finding a tree without ants wasn't easy. So on Christmas Eve we sent Daniel out hunting for a suitable one. Eventually he came up with the goods — a young tree that was growing in an awkward place, too close to one of the staff huts. After potting the tree, Audrey and Oliver began decorating it. First Oliver painted the galls in different colours, to look like bells, then they covered it with more traditional decorations. Audrey also made Oliver a large Christmas sock to put by the chimney. Oliver was very excited about Christmas. Audrey had helped him write his letter to Santa Claus, which included a list of presents he was hoping for. Then, taking our advice, he put out some mince pies and a Tusker beer for Santa and some carrots and grass for Rudolph by the chimney. He went to bed early that evening — so that the morning would come sooner. As soon as he was sound asleep, I gobbled down the mince pies, sprinkled the grass around the room and bit off pieces of carrot, which I left on the floor near the chimney. Then Audrey and I both relaxed by the fire and shared Santa's beer.

When dawn did break, we were awoken by Oliver clambering across our bed announcing, 'Mum, Dad, it's CHRISTMAS!'

'Oliver, it's 6.30 in the morning, now get back to bed,' was Audrey's less than festive response.

Oliver snuggled up in bed with us. An hour later his little voice was asking meekly, 'Can I open my presents now, please?'

'Oli, go and see if Santa's been,' suggested Audrey, still hoping for another hour's sleep.

Oliver leapt out of bed. A few seconds later he came rushing back, 'Mum, Dad, look what Santa's brought me!' He clambered across our bed again carrying a large parcel. Not only had Santa read his letter, but he'd generously left him what was top of his list — an Action Man jeep. It was almost eight o'clock and I got out of bed to make Audrey a cup of tea. 'Look what a messy eater that Rudolph is,' I shouted to Oliver from the other room.

It did seem a little strange spending Christmas in a hot, dry environment, but we soon got into the mood. Later that afternoon Simon and his family joined us for our traditional lunch of roast turkey with all the trimmings.

Part Three

Repatriation

9

End of a Nightmare

Although Audrey and Oliver adapted well to their new life in Africa, living in the reserve sometimes left us feeling a little isolated. There were three telephones on Ol Pejeta, one at the ranch office, one at the Lodge and another at the Tented Camp. Much of the time these weren't working, usually as a result of elephants rubbing their backsides against the telephone posts, causing them to fall over and the wires to break. On top of that, the locals outside the reserve would often steal the telephone wires and use them to make intricate bracelets to sell to tourists. For communication purposes and also for security reasons, all the managers on the ranch were in radio contact with each other. We each had our own radio call sign. I had a hand-held Motorola radio and my call sign became 9-9 (Nine-nine). Amusingly, Audrey became 9-9 and a half. Our house at Spoonbill Dam became known as 9-9's house. In fact 9-9 soon became my local name among the staff, regardless of whether they were speaking to me over the radio or face to face. After a while it became as familiar to me as my own name, and I would react — and still do — to anyone saying, for example, 999 as the number for the emergency services, as if I had just heard my own name mentioned.

Every Saturday we would drive to Nanyuki and buy enough fresh fruit and vegetables to last the chimps a week. Audrey always enjoyed these trips, as they gave her the chance to swap life in the bush for some civilisation. Oliver was very popular in town. Kenyans truly love children, and a walk through town would often result in someone feeling compelled to pick him up and cuddle him. Audrey was given the name 'Mamma Oliver' by the local Africans. In Kenya it is a tradition, after a woman has had a child, to call her 'the mother of —' (the child's name) rather than by her own name.

Some of the sadder faces of African society were those of the many street children who wandered aimlessly around the towns and villages of Kenya begging for money or handouts. Nanyuki was no different and the familiar words, '*Mzungu* [tourist], give me shilling' or, 'Give me pen' would ring out from every corner as you strolled by. We took a liking to one of the Nanyuki street boys, John. He was always polite and smiling and would always look out for Oliver to say hello. Although we were never sure if he was merely being friendly to Oliver so that we would be more sympathetic towards him, we were glad to support him with loose change and Audrey would sometimes buy him bread or bananas, or the odd item of clothing. In general, though, it was not wise to give the street kids money, as the temptation among them to take drugs was strong.

We weren't sure if John really was an orphan or if his parents were sending him out each day to beg for money. Then, early one morning, I saw him with a plastic bag fixed firmly to his face. He was sniffing glue. I walked over to him in the street and told him off, saying that if he was going to kill himself, then we would stop giving him money. He promised me he would stop, and in fairness I never saw him taking drugs again.

The British Army had a barracks in Nanyuki, and you always knew when the soldiers were in town, because the female population trebled overnight. Word soon spread around Kenya that the notorious British squaddies were in town, and prostitutes would flock to the many cheap hotels and seedy nightclubs, a lot of them coming from as far as Mombasa, more than five hours' drive away, to ply mankind's oldest trade. But it was worth the effort as the soldiers were a ready and willing market. This despite the real threat of catching AIDS: it is estimated that in Kenya around two million people, 13 per cent of the population, carry the HIV virus.

Meanwhile, back at Spoonbill, our garden was developing well. All Audrey's hard work seemed about to reap its reward. We were hoping for some bananas soon and small papayas were taking shape on the trees. There were several large marrows and pumpkins in our vegetable garden, and I was looking forward to eating one of Audrey's specialities for dinner — stuffed marrow. She is an accomplished cook, and her repertoire includes several Mauritian dishes, one of which is made using pumpkin leaves and another with beetroot leaves, both cooked in the same way as spinach and with added onion, garlic and ginger.

Then one morning we awoke to find that everything had changed overnight. As I stepped outside my front door, it was as if a bomb had been dropped on our garden, and yet we hadn't heard the blast. I surveyed the devastation; there was nothing left of anything we had planted. The banana, tree tomato and papaya trees had been champed down to the ground by a very large animal. All that remained of our garden were a few indigenous bushes and, surprisingly, our lone *Maerua triphylla* tree, a short tree with small, white flowers that attract many insects. The leaves are heavily browsed by wild game, and elephants are said to love them. Obviously our night watchman had fallen asleep. I found the staff outside at the back of the kitchen.

'Pole sana,' Swahili for 'very sorry', they said.

'It was an elephant,' added Grace, looking daggers at a sheepish Steven. I went to look at the vegetable garden and found that the back fence had been ripped open. Everything in the garden had been eaten except our chilli bushes — evidently African elephants don't like spicy food. It was very depressing, as all our hard work had been for nothing.

A few days later there was another strange occurrence. A clear trail of large, brown elephant footprints appeared on our green lawn. For whatever reason — maybe because the grass was cut very short and there had been an overnight frost — the elephant had killed the grass everywhere it had walked. If ever there had been any doubt which animal had attacked our garden, there was none now. These incriminating footprints showed the point where the elephant had entered and his subsequent meander through our garden and vegetable plot on his greedy trail of destruction. Now that we had our evidence, all that remained was to line up the 100 or so elephants in the reserve and get each in turn to place a foot on one of these footprints. Well, it worked for Cinderella!

After the incident with the elephant we decided to swap Steven and Daniel around, so that Steven became the gardener and Daniel the night guard. Grace warned us that the elephant would no doubt return the following evening. She added that pumpkins to elephants are like sweets to children. She was right about this, for that same evening a large, lone, male elephant rumbled up to the perimeter of our fence at around ten o'clock. Our garden must have seemed like an oasis to all the hungry animals. He began tugging at our electric fence with one of his tusks and we were alerted to his presence by the twang of wire being ripped off the insulators on the posts. Conveniently for the intruder, an elephant's tusks are completely insulated against electricity.

Leaving Oliver safely asleep indoors, we leapt into action and ran outside. We found the staff already there on the front line and joined the battle, throwing sticks and exploding lumps of soil to scare him off. The elephant was clearly irritated at this disruption to his culinary plans.

It was actually quite scary and potentially dangerous, as in the pitch dark all we could see was a huge, angry silhouette. He would retreat into the darkness, only to return a few moments later to attack from a different direction. He seemed determined to double-check for anything edible he may have overlooked last time.

We soon discovered that the short electric fence surrounding our compound provided only meagre protection against the larger animals, such as elephants and buffalo. The smaller animals could come in at will, as the fence consisted of just two strands of wire. Not long after our garden was trashed, one of our cats had a close encounter of the feline kind. We had just finished having lunch and I was walking to the kitchen, when, to my amazement, I saw a female waterbuck rush through our back gate, by the dam, then diagonally across our garden past our cat, Tinge, and out through the front gate. Tinge was patiently sitting in front of a small rodent's hole and never noticed the waterbuck pass within a few yards of him. This was very strange behaviour, as waterbuck are among the most nervous of antelopes in the reserve and they never came close to our garden in daylight. So my first thought was that something must be chasing her. Sure enough, a few moments later a beautiful lioness trotted through our back gate and boldly headed off in the same direction as its prey. The waterbuck had presumably seen man as the lesser of two evils.

The lioness suddenly stopped when she came across a bite-sized morsel in her path. She was less than ten yards away from Tinge as she crouched down and prepared to attack. With his nine lives rapidly ticking away and nature's theatre of life performing all around him, Tinge continued to sit motionless. He was concentrating all his efforts on catching his own midday snack, totally oblivious of the fact that he was about to become one himself. Acting like a lion himself, he seemed about to be eaten by one. Fortunately, I had seen what was happening and quickly called out his name. The lioness, who was bracing herself to pounce, turned in a flash and glared at me. Tinge, on hearing his name, also looked up. Thinking it must be dinner time, he decided to give up on his elusive prey and trotted straight into the house. He never even saw the lioness. So much for a cat's sixth sense!

The lioness, frustrated at now losing both meals, arched her back, and with a sideways movement of her head, snarled at me. She seemed genuinely angry. Unperturbed by my presence, she sat down watching me. A few minutes later she was joined by four other lionesses. They stayed in our garden for about half an hour, lying beneath the shade of our *Maerua* tree and occasionally surveying the house. The waterbuck's unusual act of desperation had undoubtedly saved her life.

There were over 40 lions in the reserve and they needed to be treated with the utmost respect. Our house staff could testify to this as they had had their own close encounters with lions — probably the very same individuals that almost ate our cat. During the day Steven had spotted several vultures soaring high in the sky over the garden. This meant there was a recently killed carcass in the close vicinity. Our staff watched where they landed and decided to see if they could grab some left-over bush meat. They found the kill, an impala, but also unexpectedly found the killers, a pride of six lions, resting close by in the shade. Steven and Daniel clapped their hands, expecting the lions to run off. The females did, but the male hadn't read the script, and tore after them, roaring furiously. The terrified staff screamed in unison and fled ignominiously to our kitchen. In that instance the lion had reaffirmed his status as king of the jungle, while man was relegated to nothing more than a food item on the royal menu.

As a result of all these episodes I became acutely aware of just how vulnerable I was when turning off the generator in the evenings, as I had a 30-yard walk back to the safety of the house in the pitch darkness. As I turned off the generator, the heavy, clanking noise gradually died away, leaving an eerie silence. If an animal the size of an elephant could silently enter my garden at night and lions could boldly wander through in broad daylight, then either could easily be in my garden watching me right now. With my eyes not adjusted to the darkness and armed with only a torch, I would scan the darkness for any large, predatory eyes reflecting back at me or a wayward trunk heading in my direction. During my whole stay at Sweetwaters I was constantly vigilant and on the lookout for danger.

Undeterred by the setback to our garden, we began repairing the fence and planting more seeds. One consolation was that the climate and fertile soil enabled things to grow very rapidly, and within three months we had another healthy crop of vegetables waiting to be eaten by us and not by elephants.

For Oliver, the reserve was a constant education, and we didn't have to go far to find interesting wildlife. In the evenings we would stroll around the garden together with a torch held to our foreheads, like a miner's helmet, and scan the grass. Everywhere tiny diamonds would instantly light up like dew in the grass. On closer inspection, however, we would discover that these sparkling gems were in fact the eyes of numerous spiders.

Another favourite pastime of Oliver's was to lie on the grass in front of a small, round hole and wiggle a grass stem inside to tempt the occupants to the surface. These neat, shallow burrows belonged to baboon spiders, a species of tarantula. They would attack the invading grass stem and Oliver would try to fish them up to the entrance of their hole to see how big they were. Up to six inches in size, this large, hairy, dark brown and highly aggressive spider would attack the grass, rearing up to reveal venomous half-inch fangs on its front legs. Baboon spiders are solitary, terrestrial and nocturnal, and will eat any small mammal, reptile, amphibian or invertebrate that they can overpower. Our garden was full of their burrows. Occasionally in the evening we would come across one as it walked over our lawn. The females are very impressive, as they are much bigger than the males. Not only does the female live almost twice as long as the male — around 20 years — but she also has a significant bearing on his longevity, as after mating she usually tries to eat him. However, the male has evolved a special adaptation to prevent this: a pair of mating hooks on his front legs, separate from the fangs, which he uses to hold down the female's deadly fangs during mating.

But, of course, learning about wildlife wasn't going to be sufficient for Oliver. He needed to go to school, where he could both gain a conventional education and mix with other children. Sadly, there were no schools of a high enough standard in the area — that is, ones that taught the curriculum he would have followed in England. We had often discussed getting together with the local community and creating a school on the reserve. Most people were enthusiastic, but in the end, apart from talking, little was ever accomplished. So we brought some learning books over from England, and each day Oliver would spend an hour with Audrey, learning to read and write and doing elementary maths. This wasn't really enough, though, and we grew concerned that his education would suffer.

None the less, Oliver's vocabulary was increasing fast, and naturally included many animal names. But for some reason he couldn't say the word water, and would instead say 'yoyo'. Oddly enough, this difficulty even extended to words with the word 'water' in it. So if he saw a waterbuck he would call it a 'yoyobuck'. He was able to say 'warthog', but had it been 'waterhog' then he would have said 'yoyohog'. It was all very sweet and when he did eventually learn to say 'water' we were both rather sad as it was a sign he was growing out of his toddler stage.

* * *

Sophie landed safely at Nairobi airport on 13 January 1994. It had been almost eight months since I'd last seen her. I wondered if she might ignore me — a recognised phenomenon with infant chimps separated for long periods from their parents or foster parents. Would she remember me even? I saw the crate for the first time. The KWS vet, Elizabeth Wambwa, was standing next to it, waiting to sign the official paperwork. As I approached I could make out Sophie's eye peering through a small hole in her wooden crate. She saw me approaching in the distance and immediately whimpered. At least she recognised me, I thought. She must have been terribly scared and confused. She had been shut in the crate for two days and I doubt if she had managed to sleep much at all. I knelt down and tried to reassure her. The crate made a thumping noise as Sophie began rocking with her back banging against the side. I suppose if she could have wished for anyone to be there after such a long journey, it would have been me. It must have been a shock for her to see me there after all this time.

She was held at Customs for three hours while we waited for various documents to be authorised and for relevant people to come back from their extended lunch break. But finally all the formalities were taken care of, and we carefully loaded the crate on to the back of a pick-up and set off for Sophie's new home.

The next day I was informed that Richard Leakey had resigned as director of the Kenya Wildlife Service. Sophie had been extremely lucky, as there was no telling if the new director would have allowed her into the country. Leakey had decided to go into politics, and soon afterwards he was to form an opposition party called Safina. The party was launched before the general elections in 1995, amid a fanfare of publicity. But this backfired under somewhat humiliating circumstances when the Kanu-dominated government refused Safina official registration. In fact the party didn't achieve this until late 1997.

Leakey was replaced as director of the KWS by Dr David Western, a conservationist, born and raised in Kenya. Before his appointment the KWS had agreed to let Sophie be quarantined for one month inside the unfinished chimp building. There was a buzz of excitement among the staff, as they had never seen a chimpanzee before. Sophie was understandably nervous of all the commotion and initially wouldn't come out of the crate. When she finally emerged I was shocked to see how pale she looked. Her pallid complexion made her lips stand out as if she were wearing black lipstick. It wasn't surprising, as she had been living inside the chimp house at Chester for over a year without any exposure to natural sunlight. But otherwise she looked in good health. She jumped into my arms, her lips drooping into a sorrowful pout, and clung on tightly to me with her arms and legs wrapped around me. She gave me an open-mouthed kiss on my neck and held it there for several seconds. I had missed her kisses and it was a tender moment. Then she scrambled to the ground and grabbed my arm, pulling me along and wanting me to chase her. She had been out of her crate for just a few minutes and yet she looked so happy and already wanted to play.

I left her alone for a short while and drove to the house to collect Audrey and Oliver. They were both very excited, as they had seen Sophie only once since she left Meadowtown almost two years earlier. It must have been a very happy moment in Sophie's life to be reunited with all three of us. She gave both Audrey and Oliver in turn a long hug. She then dragged Oliver by the arm and started playing with him, as she had done with me earlier. I can't imagine the emotions that must have been going through her head. After all the pain and suffering she had been through at Chester, and the long flight, which must have been both frightening and tiring, she was now reunited not only with me but with her entire family.

That Sophie had endured the endless beatings, the broken bones, the partly amputated toe and the many smashed teeth long enough to be rescued alive from her nightmare was, I suppose, partly down to her strength of character and ability to keep a low profile. But, then again, had she not gone through this torment, I'm sure the zoo would never have allowed her to follow me to Africa. That night I slept with Sophie on a fold-up bed in the chimp building. She had a restless night and we both slept very little. But I greatly enjoyed my sleepless night.

Sophie wasn't too pleased to be confined inside while the building contractors added the finishing touches to the roof. But at least she was let out of her cage at regular intervals and allowed the freedom to run around inside the building. After three weeks our local vet, John Richardson, anaesthetised her to take a blood sample. We took the opportunity to weigh her — she was now 38½ pounds (17.5 kilograms). Some days later John rang me with the results, and she was given a clean bill of health.

Now that Sophie's quarantine was over, I carried her into the car and set off for the house. Spoonbill Dam was to become her home for the next two years. Now she was truly free. She sat on my lap and looked out of the window as we drove though the mud — it had been a long time since we had been together in a car on those long journeys from Meadowtown to Chester. Audrey and Oliver were there to greet her, along with our smiling but slightly anxious staff. I don't know what she must have thought when she saw our cats again; the very same pesky ones she had enjoyed chasing a year and a half ago. She instantly recognised them and was pleased to see the mother, Sabby, whom she had always liked (even if the feeling wasn't mutual). But first she had unfinished business to attend to, and immediately set off chasing the two males around the garden. It was good to see that some things hadn't changed. After a short chase, she turned to Sabby and cornered her by the front door. She picked her up and sat there for a while, holding her tightly in her arms. Sabby grimaced in revulsion as Sophie cuddled her. It reminded me of a scene from the Walt Disney cartoon *Pepe Le Pu*, where an amorous, but pungent, skunk mistakes a domestic cat for a potential mate.

Sophie was bad news for the cats. They had been enjoying their lives in the African bush, but now they had to be on their guard all the time. Chasing the males was a boost to Sophie's ego, and if she couldn't be bothered to chase them, she would simply shoo them away by waving her arm towards them and stamping her foot. Tufty wisely spent most of the day sleeping in a cupboard in the kitchen, emerging only when it felt completely safe to do so. Sophie found scaring Tufty into the cupboard tremendous fun and would pursue him into the kitchen, slamming the cupboard door behind him as hard and as loudly as she could. Even as a baby, Oliver never paid much attention to the cats. I suppose they weren't that interesting compared with Sophie. As for the cats, they never fully trusted either Sophie or Oliver. They were used to people approaching them calmly, with a soft voice, and stroking them, and never quite came to terms with small children and chimpanzees rushing up and unceremoniously grabbing them. Oliver would try to be friendly but eventually lost interest in his four-legged friends.

For the first few days we let Sophie sleep with us in the house. We arranged a small bed for her in our spare bathroom. Later we converted our outside store room, near the kitchen, into a home where she could sleep at night. She wasn't happy with this arrangement and wanted to sleep with us. But after a while she begrudgingly accepted it as her bedroom, and in the evenings I would carry her around on my back until she grew tired and allowed me to put her to bed.

⋆ ⋆ ⋆

There wasn't much in terms of entertainment at Sweetwaters. We would often treat ourselves to Sunday buffet at the Tented Camp. The food was excellent, and it allowed us to have a change of scenery and a chat with the hotel manager, Ali Shea, or maybe meet a few interesting tourists. Nanyuki had little to offer either, other than a couple of sleazy nightclubs. So we would create our own entertainment. One of our favourite pastimes was to go on a night game drive through the reserve — a mini-adventure because you never knew quite what was in store. We had a powerful spotlight that plugged into the car's cigarette lighter. There were certain nocturnal birds and animals you would be guaranteed to see, including nightjars, bushbabies and jackals. But if you were lucky, then you might come across lions and other night-time predators, such as spotted and striped hyenas or maybe even a bat hawk. Leopard or aardvarks were the ultimate goal in night game viewing; most people living in Africa spend their entire life without ever seeing the elusive aardvark.

There were many leopards in the reserve, but the chances of ever seeing one during the day were extremely remote. One evening as we were driving past the marsh — a large swampy area in the reserve that attracted many grazing animals and predators — we discovered a leopard sitting by a bush, minding its own business. As I scanned the area with my spotlight a hippo suddenly came into view in the distance, ambling along on the ground. Hippos often come on to land to graze on grass, but generally you see them only in the water. The hippo was heading in the direction of the bush, unaware of the leopard sitting behind. They were on a collision course and each was completely oblivious of the other's presence. The hippo took one step in front of the bush, no more than a foot away from the leopard's nose, and the outcome was quite comical. Both got the shock of their life, the leopard leaping backwards about five feet into the air and the hippo about a foot in the opposite direction.

One of the good points about Sweetwaters is that the animals are not habituated to the tourists. You rarely come across other tourists when out on a drive, and as a result the animals are the way nature intended them to be — savage, aggressive and truly wild. Because of the thick bush, they can also be difficult to spot, and it was more than a year before I saw my first cheetah — two magnificent males hunting

Thomson's gazelles on one of the smaller plains. And it was almost two years before I saw my first serval cat, only 100 yards or so from the chimp sanctuary. The lions, elephants, rhinos and buffalo needed to be treated with respect. Take a drive through Nakuru National Park, where you will probably have to weave your way through large herds of buffalo that just stand there in your way, like cattle chewing the cud on a narrow country lane, or watch lions in the Masai Mara sleeping under your vehicle, and you will probably understand what I'm talking about. During my short time at Sweetwaters I was chased by several elephants, a rhino and an entire herd of buffalo.

For Sophie, our night game drives were a special treat. They also meant that she could stay up later with her family. She would pay just as much attention to the animals as we did and enjoyed every minute. On one such drive we came across a large herd of buffalo and stopped the car next to them. Sophie, who was sitting in the back, immediately wound down her window and hooted at them, banging her fist on the side of the car and waving her arm at them. Normally the buffalo at Sweetwaters are bold and aggressive. But in this instance they were so shocked that they immediately wheeled around and headed off in a cloud of dust. Feeling very satisfied with herself, Sophie wound up the window, sat back in her chair and we continued our drive.

With the chimp sanctuary almost finished — all that remained to be done was to paint the outside of the building — we needed to prepare for the arrival of the first occupants. Kenya already had a few adult chimps that needed suitable homes, but not enough to be a tourist attraction. The KWS had four adults at their orphanage in Nairobi, and there were three others at the home of Karl Ammann, a Swiss wildlife photographer who lived in Nanyuki. Sadly, two of the chimps at the KWS orphanage died before we could bring them to Sweetwaters. We decided that Karl's two adult chimps, Charlie and Judy, should be the first to arrive. He had decided to keep his other chimp, Mzee, at his large house on the expensive side of town, an area nicknamed 'Millionaire's Row', where his neighbours included Stephanie Powers and John Hurt. Karl also had a pet cheetah, which he kept on the balcony of his bedroom. Before bringing the two chimps to Sweetwaters, Karl thought it would be a good idea if I went inside their compound so I could get used to them and them to me. So I began going in with them regularly with Karl. Charlie had previously attacked an Australian girl, when he was startled by a helicopter hovering above. So already I knew he was less than predictable.

One day Karl told me that he was too busy with work and that I would have to go in alone with the chimps. He assured me I would be fine. Although I felt decidedly uncomfortable about this, I agreed to go along with the idea. I entered the compound and sat on a log. Judy approached me and I began grooming her face. She had suffered polio as a baby and was crippled in one leg, which was stunted. She couldn't walk properly but, despite her handicap, was a strong and confident chimp. Charlie then came and sat next to me and started grooming my foot. Karl had found Charlie for sale in Kinshasa, and feeling pity, paid a nominal amount for him and brought him back to Kenya. Things seemed to be going smoothly until Charlie suddenly grabbed my ankle and sank his massive canines into me. He yanked my left foot up, sending me landing heavily on my back. His hair was erect and he was going berserk, leaping around, while continuing to bite deeper and deeper into my ankle for about 20 seconds. The pain was excruciating and I could immediately feel my hot blood gushing down my leg.

'Oh my God! Help!' I screamed, but no one could hear me.

I tried to put my hand in Charlie's mouth to stop him biting me but only succeeded in getting my finger bitten too. I was powerless to stop him and I felt sure I was going to die. Meanwhile a surprised Judy, who had been enjoying being pampered by me, was suddenly faced with three choices: ignore what was going on; join in the attack on me; or risk injury to herself by supporting me and attacking a larger and much stronger male. Fortunately, she chose the third option. She leapt across my horizontal body and launched into a ferocious attack on Charlie. A big fight ensued, with much screaming.

During the fracas Charlie let go of my foot and I was able to crawl away to safety. Judy emerged victorious and unharmed. Karl, who had not heard my screams, did, however, hear Charlie's and came rushing to investigate. He found me hopping out of the gate on one leg. I was scared to examine my wounds, fearing a terrible mess. Fortunately, Charlie had not attempted any ripping movements while biting me. He had continued biting deeply in the same place, leaving me with four large gaping puncture wounds. But I had lost over a shoe full of blood, and Karl drove me to the local hospital. There a doctor cleaned the wounds but decided not to use stitches in case they became infected. He informed me that the bite wounds most likely to become infected were from humans, because of the bacteria in the saliva. I told him that as chimps are 98.4 per cent genetically similar to humans, it didn't augur well for me.

In fact by the time I reached home, my leg had swelled up to my knee so that I couldn't wear my shoe. I was unable to walk for several days. I was in such pain that for the first two days I was confined to bed. Even getting up to go to the toilet made me feel sick. The swelling and the pain lasted for two weeks. Why Charlie attacked me I will never know for sure. Possibly he was jealous of my grooming Judy or maybe he had smelt the scent of Sophie on my left leg, as she would often hold on to it and hitch a ride. I was left with four permanent scars where his canines had punctured me. In a way I had been very lucky, as I could easily have been killed or badly mutilated by either Charlie or both of the chimps. I was disappointed that at no point did either Karl or his wife, Kathy, apologise to me for what had happened. They seemed more concerned about Charlie's reputation being damaged than about any damage to me.

* * *

A month later Steve Hogarth came to stay with us for a week to see how Sophie was getting on. Steve was very fond of Sophie and was relieved that she had found herself a safe home away from the zoo. Sophie, though, wasn't quite so pleased to see him, and greeted him with a firm nip to the hand, presumably because she associated him with her unhappy period at Chester and possibly thought he was coming to take her back to the zoo. During this time we were visited by Christine Manning, who was looking after the chimps at Entebbe Zoo, in Uganda. We had previously discussed on the telephone the possibilities of some of the Entebbe chimps coming to Sweetwaters as the zoo was overwhelmed with orphans. Christine was visiting friends in Kenya and wanted to see the set-up. When she arrived she placed her bag on the ground and we shook hands. While this was going on, Sophie wandered up and had a quick rummage through the bag to see what she could steal. Christine noticed this and smacked her firmly on the head, causing her to whimper loudly in pain. Sophie wasn't used to humans showing any aggression to her, especially ones she hadn't met yet.

Over a drink in the garden we talked about

chimp stuff. Throughout our conversation I could see that Sophie was in an unforgiving mood. In fact she spent the next 30 minutes slowly circling Christine in a menacing way. Her circles became smaller and smaller until she had manoeuvred herself into a position where she could attack. Both Steve and I could see that Sophie was intent on getting her revenge, but we were powerless to prevent her as she seized her opportunity while Christine had her back turned. In a flash she had rushed over and bitten her hard on the ankle, drawing blood. Her mission accomplished, Sophie disappeared from view. I apologised for her behaviour and Christine was very understanding. It had taken Sophie almost an hour but she had eventually settled her differences.

The moral of this story: never cross a chimp — they bear a grudge and have long memories and big teeth.

10

Congo Fever

Now that the construction work was finished we had an impressive sanctuary but no chimps. Jane Goodall wanted to move her 20 Burundi chimps from the war-torn capital, Bujumbura, to Sweetwaters, and Entebbe Zoo also had more chimps than it could manage. However, I was informed that many of the Entebbe chimps had tested positive for TB, so we decided not to consider any of them for Sweetwaters.

Russell Clarke sent me to Kinshasa and to Brazzaville for meetings with Zaire's Minister of the Environment and the Republic of Congo's Minister of Water, Forests and Fishing. The purpose of my mission was to discuss the possibility of providing a safe refuge at Sweetwaters for their many confiscated orphan chimpanzees.

On entering Customs at Kinshasa airport, I was quickly summoned into the Immigration Office, where the director of the airport asked me why I was visiting Zaire (as the country then was). I told him I had a meeting with the Minister of the Environment. He looked me up and down for a brief moment and threw my passport into his drawer. Then, inexplicably, he rudely told me to get out of his office. I went out and sat on a wooden bench. Much later he emerged from his office, locked the door and left the airport, apparently, as I was informed, to go home for the night. I waited and waited for him to come back, but he never did. The hours went by and it was now dark outside. I had been waiting eight hours when Graziella Cottman, the manager of the chimp orphanage at Brazzaville, managed to get permission to enter the airport and found me sitting outside the office.

By this time I was convinced I was going to have to spend the night on the hard bench. I explained to Graziella that the director was keeping me a virtual prisoner. No doubt he was hoping to soften me up for a bribe the following morning. Graziella had crossed the Congo River from Brazzaville to meet me at the airport and had herself been waiting all this time for me. However, she eventually managed to get word of my plight to the Minister of the Environment, who immediately sent his personal secretary and his son, a colonel in the army, along with a letter ordering my release.

The director was summoned from his home and arrived 20 minutes later. We went into his office, whereupon the letter was handed to him. Looking a shade uncomfortable, he stamped my passport and, avoiding my gaze, handed it back to me. I looked him squarely in the eye, and sarcastically thanked him very much for his help.

The Minister, Mr Ramazani, was very apologetic and embarrassed about the treatment I had received as a guest in his country. As it was now rather late, we agreed to meet the following morning. After being held for nine hours I was terribly thirsty. I managed to buy a bottle of Coca-Cola from a store, but it cost me a staggering 19 million zaires! This absurd currency was the result of the hyperinflation that had been strangling this potentially rich country for many years. In real terms, it wasn't a lot of money — around two Kenyan dollars — but all my Zairean currency was in one-million-zaire notes and it seemed like a lot of notes to be paying out for one drink. Still, it felt good to be a millionaire for a day.

The following morning I finally had my meeting with the Minister, and also with Mr Mankoto, head of the Institut Zaïrois pour la Conservation de la Nature (IZCN). The trade in bush meat and orphan chimps was a major problem. Although the IZCN had the will to confiscate illegally detained animals, they didn't have adequate housing facilities or the funds to maintain them afterwards. We discussed the prospects of Sweetwaters being a solution to their problem, and they sounded genuinely interested.

After the meeting, I met Delfi Messinger, an American woman who was devoting all her time to caring for more than a dozen orphaned bonobos. She was hoping to find a safe home for them in a European zoo, and I admired her courage and dedication. We briefly chatted about the possibility of the bonobos coming to Sweetwaters on a temporary basis until she could find a permanent home for them. Graziella was with me and we both visited Delfi's bonobos.

Bonobos, or pygmy chimpanzees, are a separate species from the common chimpanzee. The names 'pygmy' and 'common' are misnomers, because bonobos are roughly the same size as the smallest subspecies of common chimpanzee, and common chimpanzee are by no means abundant. Genetically, bonobos and common chimps differ by only 0.7 per cent, and the two species are identical in their genetic relationship with man. Endemic to Zaire, bonobos are more slender and longer-legged than common chimps. They have smaller heads, and adults resemble juvenile common chimps. Their faces are black, with pink lips — they are known locally as black-faced chimps — and the hair on their head has a distinctive parting in the middle.

Despite their genetic closeness, the two species of chimpanzee differ in many ways, notably in their reproductive biology and their promiscuous sexual behaviour. Female bonobos are sexually receptive almost continuously, and both males and females, regardless of their age, regularly have sex, not just to reproduce but also for pleasure. Bonobos have similar birth rates to common chimps, giving birth at intervals of between five and six years. The sexual act can be between two partners — male and female, infant and adult, and can be heterosexual or homosexual — or it can be a group activity. Masturbation is also common. Copulation takes place in a variety of positions, including face to face, which in common chimps is very rare.

Unlike common chimps, the female bonobo's genitalia — vulva and clitoris — are orientated forwards on the body. Sex often occurs at times of tension and excitement, for example after aggressive encounters, during reconciliations or when a fruiting tree is found. Common chimps, too, will briefly mount, and touch the genitals of, other members of the group in similar circumstances, especially when two males meet after a period of separation. Bonobos also perform sex while resting — seemingly just for the fun of it.

The bonobo's sexual behaviour is often said to resemble that of humans. But my own belief is that it bears more resemblance to grooming in common chimps, and although bonobos do frequently groom each other, they appear to be using sex as a social tool, in much the same way as common chimps perform social grooming at times of excitement and tension. Perhaps it would be more accurate to describe their behaviour as 'sexual grooming'. Bonobos are less aggressive than common chimps, and females tend to dominate males. Their vocalisations are also distinct from those of common chimps, with sounds that are much higher-pitched, especially when young.

After lunch with Delfi, Graziella and I crossed the Congo River by boat to Brazzaville, on the other side of the river. Everyone crammed on to the small open boat and there didn't seem to be any order to the proceedings. The water looked peaceful and still, belying the speed of the current below. We each had a wooden seat. At the very last minute a woman was carried on to the boat on a stretcher. We all shuffled together, trying to make room for the stretcher. The woman was in a lot of pain and every now and then groaned in agony. Her companion seemed very anxious and tentative, all the time wiping the sweat that was pouring from her brow. The boat glided across the still water and some 20 minutes later docked in Brazzaville. After my passport was stamped at Customs we walked outside to try to find a taxi. This didn't take long and soon we were hurtling along to Brazzaville Zoo. It felt great to be back. I had fond memories of this country. At the zoo I met the Attwaters and Jean Mboto again. I was pleased to see that Grégoire was still alive and well. Even the lions were still there — as skinny as ever, and living off air, it seemed.

A couple of days later we travelled down to the Jane Goodall Institute Sanctuary at Pointe-Noire. Paul Actel had resigned from his job to travel around Africa. We met up with Geoff Creswell and his girlfriend, Liz Pearson, who were now running the sanctuary. La Vieille recognised me instantly as that horrible man who had stuck a dart into her backside. She screamed when she saw me and turned around and presented her bum in a submissive gesture.

After a couple of days I was struggling to find any energy to walk for long periods in the sun and was surprised by how much the heat had sapped my strength. Later that evening I noticed that I had a slight red, blotchy rash, which was beginning to itch near the inside of my right elbow. I didn't take much notice of it at first, but by the following morning it extended halfway down my lower arm. I looked up the symptoms in David Werner's *Where There's No Doctor*, an excellent book and a must for anyone planning a long trip abroad. The symptoms all pointed to typhus or tick fever. I remembered finding a large tick on that very area about ten days before, when I was still in Kenya. At the time, the tick was engorged on my blood and had possibly been there quite a while — it is only after you remove them that they begin to itch. We decided that I should return to Brazzaville as soon as possible. Geoff said he would drive me to the airport the following morning.

By the time I reached Brazzaville the swelling had reached the middle of my upper arm and I had developed a high fever. The next day the rash was up to my shoulder and halfway across my upper back and the right side of my chest. Graziella called in Dr Ernest Davies, the local American doctor, and he took a blood sample. The next day the rash had spread down to my fingers and my right hand was twice the size of my left. I could see that the doctor was panicking as he wasn't 100 per cent sure what I was suffering from. The blood test confirmed it was typhus, often known as tick fever.

On the third day Dr Davies told me that if there was no improvement by the next morning he would have me evacuated back to Kenya. The fact that he seemed so anxious didn't make me feel any better, and that evening I lay in bed wondering if I was going to die here in

Brazzaville. Fortunately, by the morning my fever had broken. The swelling in my arm was also beginning to subside. I stayed in bed for a further two days with a temperature. But I had no time to feel sorry for myself, as I had a meeting scheduled for the following morning with Rigobert Ngovlali, the Minister of Water, Forests and Fishing. So, despite still having a temperature of 102 degrees, I dragged my body out of bed and attended the meeting. I felt quite nauseous during our discussions but managed somehow to get my point across.

The general conclusion of both my meetings was that they agreed with the principle of allowing orphaned chimpanzees to be expatriated to Kenya but at the same time wanted to know what we would be able to provide for their countries in return.

The following evening I accompanied Graziella as she drove some journalist friends of hers back to their hotel. There wasn't enough room in the cab of her pick-up, so I volunteered to stand on the flatbed. I still had a temperature and welcomed the cool breeze on my face. There were soldiers everywhere and many of the roads were blocked off. For some reason Graziella inadvertently turned down a wrong road, which was blocked off by soldiers. They immediately ordered her to stop, but she seemed to panic and instead decided to pretend she hadn't heard anything and continued to drive away. The nearest soldier to us again shouted for us to stop and when he saw that we were ignoring his command, he raised his Kalashnikov rifle to his shoulder and aimed it directly at me. Immediately I pounded on the roof of the pick-up with all my force, shouting at her to stop, which fortunately she did.

The soldiers walked slowly over to us and demanded to see our IDs. This was the one day when I wasn't carrying my passport; I had left it in my suitcase in Graziella's house. At this the soldiers became aggressive and insisted on arresting me. I was still not fully recovered from my illness and the prospect of being thrown into a Central African jail didn't bear contemplating. Thankfully, Graziella convinced them that it was a genuine mistake on my part and after we'd negotiated my worth with a fistful of dollars we paid the bribe and continued to the hotel.

The next morning I had to get up early and return to Kinshasa to catch my plane back to Kenya. As I was still weak, Dr Davies had arranged for me to be met on the Kinshasa side of the river by someone from the US Embassy. The man was supposed to assist me in getting through all the hassles of Zairean Customs and drive me to the airport. As the boat drew in to dock, a man began pointing at me and then to himself and waving enthusiastically. I was the only white person on the boat. I got off and he immediately approached me.

'Are you the man from the Embassy?' I asked.

'Yes. Let me help you. My name is Christian.'

Thank goodness, I thought.

He immediately took my baggage and I followed him through Customs. We were ushered into a room where an official inspected my luggage. In the process he came across my Canon Eos 100 camera.

'You have a nice camera. Where is the receipt for this?' he asked.

The camera was four years old at the time and, of course, I didn't have the receipt on me. He said that in that case I would have to pay $100. I told him this was ridiculous, but in the end agreed on a compromise of $30. Irritated, I followed Christian a little further down the corridor until we were ushered into yet another room, this time by three police officers. They too looked through my bags. They said that I would have to pay them $50.

'What for?' I replied.

'This is Africa, but if you want we can discuss this at the police station,' came the veiled threat. I gave them another $30, grumbling that this was all I had.

Then, just a few yards further on, I was stopped by four soldiers, one of them a woman.

'We are hungry. What have you got for us?' the woman demanded.

They were about to go through my suitcase, when I produced a $10 note. Unfortunately, I didn't have any smaller change, otherwise I would have offered them a dollar each. With a sigh of relief I emerged from Customs, slightly broke but not broken.

'I thought you were supposed to help me get through Customs without all this hassle,' I said sarcastically to my escort. I was now becoming suspicious as to whether this man really was from the Embassy.

He shrugged his shoulders and we continued to the car park.

'Where is your car?' I asked.

The man looked at me bemused.

'Do you mean to say that you have lied to me all this time and that you're not from the Embassy?' I said. At this the man looked to the ground, embarrassed.

I quickly scanned the area to see if I could see the genuine man who was supposed to meet me. But it was hopeless, as the port was teeming with hundreds of people, and I wasn't about to go back through Customs again to look for him.

'Thanks to you,' I continued angrily, 'I have now missed my lift to the airport.'

Then he had the audacity to ask me for some money for his help. I told him where he could go to get his money and clambered into the first available taxi. As we sped away, the driver told me how difficult life was in Kinshasa because there was no fuel in the city. He wasn't lying, for everywhere along the way there were people selling petrol in five-litre plastic jerrycans. Then he stung me for $50 instead of the usual $8. Now I had virtually no money left.

As I entered Kinshasa airport I surveyed the area for the director. I was worried that he might recognise me and want to get his own back for the humiliation he had suffered. Within minutes I was approached by a man who said he would help me get through Customs quickly. He wasn't wearing any uniform, not even a badge. He asked for my plane ticket, my passport and some money to help him get me through Customs. So

I handed over my very last $20. Then he told me not to move and that he'd return in 30 minutes. I wasn't happy with this arrangement but in Kinshasa everything is completely disorganised. So, as crazy as it sounds, I put my trust in him. I waited anxiously for his return. Half an hour went by and then an hour, and there was still no sign of him. Panic began to set in. My plane was due to take off in less than an hour's time. The man had told me his name was Alfred, and that he worked in the airport baggage hall.

A soldier saw that I was looking troubled and kindly asked if I needed any assistance. I told him my story and he took me to where the man had said he worked. When we asked where Alfred might be, they said that there was no one with that name working there. My heart sank and it suddenly hit me that I had given my passport, my plane ticket home and all my money to a total stranger in plain clothes. I felt a complete fool and utterly lost, and it wasn't as if I could go to the airport director and complain. I suddenly felt very weak as my temperature began to soar under all the stress.

Then there was a tap on my shoulder. I turned around and there, smiling at me, was Alfred.

'Where were you?' he said. 'I have been looking for you all over. I told you not to move, no? Come, your plane is leaving soon. You must hurry.'

Words cannot explain how relieved I was to see this man. He had done everything he said he would. I boarded the plane, found my seat and waited impatiently for take-off. I was praying for no more last-minute hitches. I was convinced the airport director would discover that I was on the plane and order me off. It wasn't until the plane was safely in the air that I felt I was free from my Kinshasa nightmare. I sat back in my seat and for the first time in a long while tried to relax. I was feeling exhausted. I closed my eyes, and for a few hours time stood still.

An abrupt jolt as the wheels hit the tarmac propelled me from the tranquillity of my sleep. I looked out of the window at the familiar scenery and breathed a huge sigh of relief.

11

Non-human Friends

At Nairobi airport I was met by Simon and driven back to Sweetwaters. It was very comforting to be home, and I spent the rest of the day in bed. Sophie was pleased to see me back and greeted me with her trademark kiss on the neck. Dickson told me that Sophie had missed me very much. He explained that, during my absence, the staff would tease Sophie by calling out my name or asking, 'Where's Vince?' Apparently, she would rush over to the house and look through each of the windows in turn, trying to find me. They weren't doing this in any malicious way, as they all loved her; more because they were astonished by her intelligence. Although I had always known that she knew my name and hers, it was a good feeling to have tangible evidence for myself.

We got on very well with all the staff and didn't mind them living with us. They became like members of our extended family. Apart from providing us with a feeling of security, they were great company, and at no time did we feel our privacy compromised. They were all fascinated by Sophie, who would spend a lot of time at their houses, where they made her cups of tea and often shared their food with her. Sophie particularly liked Steven. He was very calm, patient and assertive with her, and she respected him. In fact he was the only one among the staff with whom she hadn't at some point lost her temper. Steven was proud of the fact that Sophie liked him best.

Grace and Daniel were more reserved, wary of Sophie's unpredictability. Grace was a very strong-willed girl, and despite this being a male-dominated society, wouldn't put up with any nonsense from the men. She was hard-working and conscientious, and with her fiery temperament, the dominant member of the three staff. I suppose the men didn't know quite how to handle her. All of them were very fond of Oliver, and as he was an only child they were good company for him too. Daniel and Steven were both married with children. Their families lived outside the reserve. Daniel had six children and would occasionally bring his youngest son over to play with Oliver. The two of them got along like a house on fire. Grace had a friend who would occasionally visit her with her small baby. Sophie was fascinated by this baby girl and wanted to cuddle her all the time. The problem was, it reached a point where Sophie loved the baby so much that she became possessive and jealous when the mother tried to take her away. So in the end we advised the mother that it would be better if next time she came alone. But I suppose it showed that Sophie's maternal feelings were strong.

At the sanctuary, we were eagerly awaiting the arrival of the first chimps. We initially accepted three adult chimpanzees from within Kenya. Towards the middle of March 1994 Charlie and Judy arrived from Karl Ammann's home, followed a week later by an adult male called Barook from the KWS orphanage in Nairobi. Barook had grown up with Judy at the orphanage, as they had been confiscated together. Then Judy left to join Charlie, leaving poor Barook living alone. He was pleased to be reunited with his friend, but aggressive towards his new rival. The introductions between the males had to be done very carefully. But we eventually managed to integrate the three of them without serious incident. This was a good experience for the chimp keepers, who would soon be embarking on a steep learning curve.

It was never intended that Sophie should become part of this adult group. It would have been a bit like rescuing her from the frying pan only to throw her into the fire. Besides, I hadn't gone to all the trouble of removing her from Chester Zoo only for her to be killed by a different group of chimpanzees in Kenya. The plan was for her to be introduced into an infant group of chimps of her own age. There were ten infants still remaining in Burundi, and they would be joining us shortly once all their CITES paperwork had been sorted out. So Sophie continued to live a contented and civilised life with her human family amid the serenity of Spoonbill Dam.

On 16 September 1994 we celebrated Sophie's fourth birthday. She was still only a youngster, but back in those depressing days at the zoo there had been many times when I doubted whether she would make it to this age. As a special treat, we bought her a bar of milk chocolate. She would receive chocolate just twice a year, on her birthday and at Christmas. She gratefully took her present and sat under the shade of a bush to eat it. Normally with food she enjoyed, she would devour it as fast as she could. But chocolate was special, and she would take her time, nibbling off tiny chunks and savouring every last morsel. Later that evening Audrey telephoned Solange in England, to wish her a happy birthday too.

⋆ ⋆ ⋆

A few days after Sophie's birthday, Grace announced out of the blue that she was leaving. She was getting married to the ranch's accountant, a Kikuyu, and was pregnant with his child. We were shocked and saddened. I didn't even know she had a boyfriend, let alone that she was expecting a baby. But at the same time we were pleased that she had found herself a good husband — our loss was his gain. It wasn't that common for Turkanas to marry outside their tribe, but I wasn't really that surprised about Grace, as she was very intelligent and had such a strong personality. We replaced her a few weeks later with Elizabeth, who was Dickson's sister-in-law. Elizabeth was quite a different character from Grace, much shyer, but just as hard-working.

Shortly afterwards Daniel also left us. He was

moved to the rhino patrol in the reserve. This was a position that held much prestige among the men. Daniel had a reputation as an excellent tracker and I had to admit his talents were being wasted working for us at the house. However, it was very sweet to see him moved to tears on his final day. We replaced Daniel with another Kikuyu, Paul Some. At the same time we employed a Somali driver, Ibrahim Ali. This was so that Audrey wasn't stuck at the house all day and could go into town without me whenever she wanted.

* * *

We were nearing the end of the dry season, and the rains were long overdue. The heat was so oppressive that we longed for the rains to come and refresh us all. The grass in the reserve was now devoid of any moisture. Bush fires were a regular occurrence, especially up on the ranch. Many of them were started by members of the local community, who sneaked into the ranch in search of wild honey. On finding a nest they would smoke out the bees, take the honey and then neglect to put the fire out afterwards. It only needed a tiny spark to ignite a bush fire that could spread with terrifying speed. I had assisted in fire fighting on the ranch on many occasions and witnessed how dangerous these fires could be. Some of the flames reached 12 feet, and the heat was overwhelming from just a few yards away. The high tannin content of the *Euclea* leaves makes them explode into flame. The hot air currents rising from the heat would send burning leaves soaring high into the air, spreading the fire, so you never knew where the next outbreak would appear.

At the end of September 1994, less than a year after Sophie's arrival, we were joined by a young female chimpanzee called Tess. She was a very affectionate, tubby little chimp. Her owner, Mrs Mohammed, lived in Nairobi. She had travelled to Kinshasa to visit her sister and come across a sorry-looking baby chimp for sale in a market. Taking pity on her, she bought the chimp for a nominal amount. Then she flew back to Nairobi with Tess, and, incredibly, was allowed to walk through Customs at the airport carrying her in her arms and with no CITES paperwork. But by the time Tess was about three years old, she had became too much of a handful for Mrs Mohammed to manage in her flat, as, among other things, she pulled off all her door handles.

At last Sophie was going to have her first non-human friend. As I drove back from Nairobi with Tess sitting next to me, I was curious to see how Sophie would react to this hairy visitor to our house. I carried Tess up to Sophie, who was sitting by the front of the garage. Sophie seemed a bit jealous of me carrying another chimp. She came over and sat next to Tess. Then she prodded her a couple of times with her finger in the chest. Tess wasn't interested in making friends at this point, and was rather confused as to her whereabouts. The two occupied their own space for most of the day until late afternoon, when Sophie broke the ice and initiated a game of chase. Soon they were laughing and by the end of the day Sophie was seen grooming her new friend.

Tess was quite headstrong at first. She was wearing a nappy when I brought her back from Nairobi. As I was going to let her sleep in the house for the first night, I decided to continue with the nappies. I managed to get her dirty one off, but she refused to let me put a clean one on her, and each time I tried she gave me a firm nip on the hand. So I gave up on the nappies, and she slept and pooed in the spare bedroom. Oliver also got on very well with Tess, and would use her like a toy horse. She was always tolerant of him and, despite his weight, would let him ride on her back like a young Tarzan.

Unfortunately, Tess brought with her a souvenir of her time in Zaire. She had tapeworm, as I discovered when she began shedding wriggling egg segments, called gravid proglottides. The doctor prescribed niclosomide tablets, but these had little effect, and it wasn't long before she had infected Sophie. As cuddly as Tess was, the sight of wriggling tapeworm segments, crammed full of eggs, coming out of her bum didn't make her any more endearing.

Barely three months later another young female chimp followed. Naika had been abandoned at the Uganda-Kenya border by an Egyptian circus that was performing in several countries in East Africa. The circus owners had two adult performing chimps. When entering a new country on their African tour they were cunningly applying for entry and exit permits at the same time, stating that they had six chimpanzees, when in fact they had only two. Nobody ever bothered to verify the number of animals they had when entering the country, so during their stay they would buy four baby chimps and leave the country with them, pretending they had entered the country with them earlier. However, believing Naika was dying of pneumonia, they abandoned her at the border. Fortunately, she was rescued by Mrs Kalpana Korea, a herbologist living in Kisumu, who nursed her back to health using herbal remedies.

I drove to Kisumu in mid-January 1995 and returned with Naika the following day. Naika was much more excitable than Tess and was hooked on tea. Each morning she would scream for her tea, gulp it down, and then scream for a second mug. She got on well with Oliver and the other two chimps, although sometimes she could be quite rough when playing with Oliver. It wasn't that she was aggressive, just overexcitable. But it was fun having the three chimps around the house, and they were great company for Oliver. It also gave Sophie the opportunity to acquaint herself, and practise her social skills, with chimpanzees slightly younger than herself, and this allowed her to regain some of her shattered confidence. The luxury of these gentle introductions to life as a chimpanzee in a group was a far cry from her previous sink-or-swim experiences at Chester. This time it was on her own terms.

Sophie had endured a traumatic time at

Chester, but being young she was resilient enough to recover and was rapidly becoming once again the cheeky chimp she had always been at Meadowtown. Crucially, she now had the support of her parent. With Sophie at the helm, the three young chimpanzees, along with Oliver, quickly formed a special friendship. They continued to live at our house, sleeping together in the converted store room. Sophie was the oldest and, as the dominant force in the group, a major influence on them all — even Oliver.

We arranged for two of the keepers, Dickson and David, to take turns to come to the house during the day in order for them and the chimps to get acquainted. They were both excellent keepers and genuinely enjoyed their jobs. Our house gradually became out of bounds for the chimps, especially during the rainy season, when their feet were always covered in mud.

Meanwhile the tapeworm was proving very resistant to drugs, and after about a month, the gravid tapeworm segments began wriggling out of Naika's backside. Now we were worried about Oliver catching the parasites. We had the tapeworm identified and it turned out to be *Bertiella studeri*, which is commonly found in monkeys, and occasionally in humans. It measures about 12 inches in length by just under half an inch in width. Primates can acquire these parasites by eating orabatid mites — non-parasitic mites closely related to ticks and found in soil, rotting wood and other decaying organic matter — that are infected with the tapeworm larvae. A different drug, Droncit, was prescribed for Naika, and this seemed more effective than the previous treatment. From then on, we de-wormed all the chimps every two months.

Chimps are susceptible to a wide range of parasites that also affect humans. In the sanctuary, they at times had problems with intestinal parasites such as tapeworm, amoebas, giardia and ascaris. They tended to keep these parasites under control, however, and if there was a flare-up we would treat them with the appropriate medicine. The tapeworm proved particularly difficult to eradicate. At Kibale National Park, in Uganda, it was found that the chimpanzees were able to rid themselves of the *Bertiella studeri* tapeworm by swallowing the leaves of a small shrub called *Aspilia mossambicensis*.

The study of self-medication in animals — termed zoopharmacognosy — is a growing area of science. One of the leading pioneers of self-medication in primates is Dr Michael Huffman, who also happens to be a really nice guy. Mike spent several days with us at Sweetwaters on a couple of occasions. I took him for a walk through the sanctuary, where we found several species of plant known to be eaten by chimps in the wild for self-medication, including *Aspilia mossambicensis* and *Vernonia amygdalana*, although I personally never saw any of the Sweetwaters chimps feeding on these leaves.

In the wild, chimps have been known to chew the bitter pith of certain plants, such as *Vernonia*, and swallow the whole leaves of other plants, such as *Aspilia*, when suffering from heavy infestations of intestinal parasites. The local WaTongwe people also use *Vernonia* in traditional medicine for a variety of ailments, such as malarial fever, schistosomiasis, amoebic dysentery and other intestinal parasites and stomach disorders. In fact the pith of *Vernonia* contains antibiotic compounds that are capable of killing intestinal parasites as well as parasites that cause schistosomiasis, malaria and leishmaniasis.

Leaf swallowing is seen mostly at the beginning of the rainy season, when parasitic worm infestation starts to increase. The leaves of *Aspilia* are unpalatable and have rough surfaces with tiny, hook-like structures called trichomes. Although *Aspilia* does not form part of the chimpanzees' normal diet, they make a big effort to forage for these plants. The leaves are carefully selected and then swallowed whole. The rough leaves appear to scrape worms off the gut wall, which then become hooked and entangled in the leaves. Interestingly, the local human population commonly use *Aspilia* plants in traditional African medicine for treating ailments such as stomach ache and coughs.

Self-medication by swallowing leaves has now been seen in at least 11 different populations of chimpanzees, as well as in bonobos and eastern lowland gorillas, with as many as 34 species of plant swallowed. chimpanzees have been found to use 13 species of medicinal plant.

⋆ ⋆ ⋆

As we neared the end of the dry season the heat intensified and the atmosphere became heavy and very dry. Being smack on the equator, I would only have to stand out in the sun for ten minutes for my skin to burn. In fact my nose seemed to be permanently burnt red like a boiled lobster. I was informed that there were many cases of skin cancer among the white Kenyans, and that we needed to protect ourselves with a strong sunblock and wear sunglasses and a hat at all times. But after a while I became too lazy to bother. The light was also exceptionally bright. They say that green eyes absorb more ultraviolet rays from the sun than either blue or brown eyes. I have green eyes, and would often have headaches as a result of eye strain.

The dry atmosphere affected Sophie too, and she would often suffer from sore feet, which would crack open deeply so that at times she could only walk bandy-legged on the sides of her feet. Elsewhere the dry season was taking its toll on the wildlife. The eland seemed to suffer more than most, becoming very thin, with their bones prominent beneath their hide. Oddly, the zebras never showed any loss of condition, even when they were starving and close to death's door: they always looked as fat and barrel-chested as ever.

Then, just when you thought the rains would never come, the heavens opened unannounced. There was no build-up, no light showers to begin with — just a torrential downpour out of the blue. They don't call it the rainy season in Africa for nothing, and when it rains, boy, does it rain! The grass instantly responded and after only a few days the new growth was visible. The rainfall could be quite isolated. There was one time when the rain was absolutely sheeting down on one side of our garden and yet the other side was bone dry. The roads of black cotton soil, which had become rock-hard and dusty during the prolonged heat of the dry season, quickly disintegrated into thick, sticky mud. I became an expert rally driver in the muddy conditions, and at times would be forced to drive for over half a mile stuck in a rut at an angle of 45 degrees. The trick was to continue driving and maintain your momentum until eventually you managed to slide or bump your way out. If you stopped the car, then the chances were your wheels would begin spinning, sinking you deeper and deeper into the mud.

Each year the Ewaso N'yiro would spill its banks and flood an area extending some 100 yards or so either side of the river. There was only one way across the river by road, and that was over a metal bridge called Elephant Bridge, which had been built by the British Army. On one occasion I decided to take a chance and drive slowly across the bridge even though it was submerged beneath three feet of raging water and I could see only the top of the side railings. There was no way of knowing if the bridge had been damaged underneath. The water was up to my car window and splashing over the bonnet. The deafening noise of the rushing water was quite terrifying. The force of the current was immense, and soon many of the banks had either disappeared or been reshaped in the space of an afternoon. I couldn't swim, so had the car been swept into the river, then I would have been doomed. But I made it across, and then had to do it all over again on the way back. Reaching the chimps became a serious problem during the rains. The marsh would also be flooded and I was forever getting stuck in the mud.

During this period we regularly had floods of water and mud coming straight through the front of our house and out through the back door. On one occasion our dining room was six inches deep in sticky mud, and somewhere beneath this dense layer of slime were our carpets. Our dam was fed by a stream from the river, and floodwater poured into it, bringing with it thousands of barbel. These fish attracted numerous wading birds to the dam, which was so full of water that it was in danger of bursting its banks. The water was cascading from the dam's outlet like a waterfall, and had soon dug a sump several feet deep into the ground. Inside this sump were hundreds of trapped barbel. I managed to put most of them back into the dam, but kept around 30 for our dinner. Later that evening we shared them with the staff. Fried in batter with a squeeze of lemon, they were delicious.

Despite all the mud, the rains were gladly welcomed by everyone, including the animals. The elephants, in particular, seemed to revel in it, especially when it was raining heavily. Sweetwaters had around 100 elephants in the reserve, and on one particular morning I managed to video 34 of them all bathing in our dam at the same time. They seemed to prefer playing in the water when it was raining to bathing during the dry season, when you'd think they would have appreciated it more. The noise they were making was incredible. They were splashing about and trumpeting as loudly as they could. Some of the youngsters were wrestling with each other, while others were completely submerged, with only their trunks protruding out of the water, like periscopes. I don't recall ever seeing animals enjoy themselves so much.

Like the elephants in our dam, our four youngsters loved the rain. Rain meant mud and lots of it, and they liked nothing better than to play in it. They would wait patiently for the downpour to stop and then rush outside to the drive. Oliver would take off all his clothes and the four of them would splash about and roll in the puddles for ages until they were caked in mud from head to toe. Afterwards Tess would grab the hosepipe and hold it over her head to wash off the mud. Sophie and Naika, though, hated water, and we had to clean them with a wet cloth. Oliver didn't have a choice in the matter and would be ordered in for a hot bath.

Although Sophie didn't enjoy water, she was very hygienic and hated accidentally stepping on faeces lying on the ground. If she did, then she would carefully wipe it off her foot either by rubbing it on the grass or by picking up a cloth or some leaves and wiping it off by hand. If she defecated on the grass we would usually take a large geranium leaf from the many plants surrounding the house, pick up the faeces in it and throw it away. I was astonished to see Sophie on many occasions do the same thing: break off a geranium leaf and pick her own faeces up with it and then carefully throw it in the bushes, probably so as not to walk in it later. I wasn't even aware that she'd been paying any attention to us. After a while we tried to discourage her from doing this, as she would invariably get her hands dirty.

Sophie also made sure Oliver was kept clean and would constantly groom him every day, especially his head. Interestingly enough, this behaviour also rubbed off on Oliver. If Audrey had a scab on her arm, then he would feel the urge to pick it off — something that his mother didn't necessarily appreciate.

⋆ ⋆ ⋆

At the beginning of January 1995 I received a phone call from a friend, Dr Nina Hahn, who was working at the Institute of Primate Research in Nairobi. She had come across an abandoned baby vervet monkey and asked if we would be willing to hand-rear her, and I agreed. The monkey was about four weeks old and had been named Annie. I wasn't sure how the three chimps would react to another primate at the house, especially as, in the wild, chimps will regularly hunt and eat monkeys. But they seemed curious about Annie, and quite gentle. Sophie, in particular, was fascinated and wanted to carry her.

At first she found it ticklish when Annie clung to her hairy chest and would laugh as only a chimp can. But I was astonished to find that Sophie seemed to want me to show her what to do. When I demonstrated to her how to hold Annie correctly she would stare intently at me with full concentration while alternating her gaze between me, Annie and her arm, which I was attempting to manipulate. This was unusual because Sophie would always pull her hand away if I tried to position it in a certain way, and this was the only time that she ever freely allowed me to do this. Sophie was very ticklish and seemed unable to tolerate the young chimp clutching on to her chest. But she improvised by carrying her on her arm. Because Annie was so small she held on to Sophie's arm with all four limbs, jockey style.

Over the next few days Sophie carried Annie on her arm everywhere she went. She slept with her and, amazingly, even fed her bits of banana from her mouth. She would bite off small morsels and then drop them in front of the grateful infant. I say amazingly, because although Sophie might begrudgingly share a morsel of carrot or some other vegetable, she would never contemplate sharing her bananas or any other of her favourite foods with anyone. Sophie had obviously noticed me breaking off small chunks for Annie. I was surprised at Sophie's acts of altruism. I was witnessing her first attempts at practising motherhood. I suppose Sophie saw Annie a little bit like a child sees a toy doll.

Maybe all that time spent carrying Sophie, giving myself backache, had finally paid off. Now I felt sure that, given the opportunity, she would one day rear her own infant. I just hoped that she would be allowed the chance to breed.

Annie quickly developed a close relationship with Sophie. Sadly, however, this turned out to be her downfall. Just over a week later Annie was following Sophie on the ground. Sophie stopped and sat down near our stock of firewood at the back of the kitchen. Annie ran straight past Sophie and right inside the wood burner. This was alight at the time and she sat in the back of it. Steven witnessed this and frantically tried to get her out. He succeeded and although she was in there for only a few seconds, this was long enough for her feet to be badly burnt. We took her to the vet, who put her on antibiotics, but despite this she lost several toes. She never recovered, and because she was so young, rapidly went downhill and died a few days later. We were all very fond of her and it was a great pity that we never got to know how her relationship with Sophie would have developed.

* * *

It was now two years since Sophie had rejoined us, and she and Oliver continued to grow at a similar pace. Although Oliver was slightly taller than Sophie, with his longer legs, Sophie's body was bigger and much stronger. So, proportionally, they were still about the same size. For two years Sophie had no front teeth because they had gradually been smashed and broken by the chimps at Chester. I was worried that maybe she had permanent damage and would never grow new ones. But in the middle of February 1995 her gums started swelling at the front and her permanent teeth began erupting. At long last she could enjoy eating apples and carrots properly. She was very proud of her new teeth and, more importantly, her new-found ability to bite. Now she could assert her dominance more forcefully, and for the next few days she would nip both Naika and Tess at the slightest excuse. Oliver was curious about her new teeth and would repeatedly ask her to let him look at them.

'Sophie, show me your teeth,' he would say, and she would willingly oblige and open wide.

Oliver was also growing his own pearly-white milk teeth. During these periods both of them would occasionally feel a little 'rundown'. So, as we had done before in Meadowtown, we kept a bottle of Calpol in the house. All medicine had to be kept safely hidden away from Sophie, though, as, given the chance, she would down the whole bottle in one.

Having managed without her front teeth for so long, Sophie knew the value of them. Previously she'd had difficulties in eating hard foods, and had first-hand experience of how painful it was to be bitten. When later a wobbly milk tooth was bothering her, Steven wanted to assist in removing it. Sophie welcomed his help and constantly pulled his hand to her open mouth, encouraging him to continue. After he succeeded in extracting it, however, she wanted her tooth back from him. But Steven wanted to hold on to it, because he knew that I would want to keep it. When he refused to give it back Sophie flew into a tantrum, screaming and chasing him all around the garden, much to the amusement of Elizabeth and Paul. This was the only time that Sophie ever lost her temper with Steven. What she wanted to do with her tooth remains a mystery. Maybe she wanted to place it under her pillow that night.

After many months of clearing political hurdles, on 20 February 1995 we eventually succeeded in bringing to Kenya ten adult chimpanzees from Jane Goodall's sanctuary in Burundi. Sweetwaters chimp sanctuary had now become home to 13 chimps, all of them with one thing in common: they were orphans whose mothers' lives had been cut short by the poacher's bullet. But these chimpanzees were the lucky ones (if you can ever call them lucky). They were the few fortunate individuals who had survived the ordeal of being captured and separated from their mothers, and very much the tip of the iceberg in comparison with the high numbers captured each year for the bush-meat trade. For every orphan captured, at least one chimp — the mother — is killed. More often than not four to five members of the group will also perish. Few of these captured infants live to be sold at market. They are often injured by stray gunshot when their mother is killed. Most usually die from disease, malnutrition and dehydration. Some are kept in the village as pets for a short time and then eaten or discarded when they become too ill or lethargic to play any more. Although these orphans (like their deceased mothers) no longer contribute to the wild gene pool, they still have a potentially important role to play. In effect they become ambassadors for their own species, highlighting the plight of apes in the wild.

Among these orphans were five males — Poco, Safari, Socrate, Gerbil and Ndaronse — and five females — Jojo, Sultana, Cheetah, Alley and Furaha. All had their own tragic story to tell. Poco had spent several years in a cage so small that he could only stand up or sit down. This cage was suspended in the air above a garage in Bujumbura as a lure for passing trade. Owing to his cramped living conditions, Poco had become one of the most bipedal chimpanzees that you could ever meet, spending much of his time walking or running upright over distances of 30 yards or more.

At the beginning of September that year, Russell Clarke paid me a visit at Sweetwaters. He told me that I was being promoted to manage the game reserve as well as the chimp sanctuary. This greatly increased my workload, although sadly not my bank balance. With my time now divided between the reserve and the chimps, I set up Sweetwaters Research Centre, and soon we had university students from all over the world studying the behavioural ecology of the wildlife within the reserve.

There were 80 staff employed in the game reserve. About 20 of these worked as rangers monitoring the welfare of the black rhinos. These rangers spent virtually their entire time living in the bush. They were brave men and excellent trackers, who could identify most animal trails at a glance. They also had the knack of being able to sense danger from afar. The best and most senior ranger in the reserve was a likeable character called William, better known by his radio call sign 0-2 (Zero-Two). He was a Turkana in his late fifties, and also happened to be Grace's uncle. William had an interesting background. A poacher turned gamekeeper, he knew everything there was to know about tracking wild game. He had endured several encounters with dangerous animals, which had given him a kind of celebrity status, and all the staff had tremendous respect for him.

William's luckiest escape was when a male lion leapt on to him from behind a bush. The lion knocked him to the ground and stood over him, biting his arm. But William managed to kill it using only a knife with his free hand. The bites had severely damaged the muscles in his arm. He had another narrow escape when he was tossed by a rhino, which left him with a dislocated shoulder. Slightly less impressive was when he was put in hospital by a warthog. William was standing over an old aardvark hole when the squatter unexpectedly rushed from the burrow, slamming through his legs and slicing them to the bone with both tusks. I was never quite sure if William was a hero or just plain accident-prone.

The dangers in the reserve were many, but you always felt safe in the hands of the rangers. Not long after I was promoted we had a report of an injured baby elephant, seen trailing its front leg, so I set off with two rangers to have a closer look. We got as far as we could with my car until we reached a stream, where we abandoned the car and continued on foot. The two rangers, one Turkana, the other Samburu, both chose to walk barefoot. The soles of their feet were as tough as old leather. The grass was full of needle-sharp acacia thorns but the rangers were totally unaware of them as they walked boldly and without caution through the thick bush. As we continued our pursuit of the herd, they informed me that we had just entered the territory of a particularly aggressive male black rhino. Several rhino middens marked its territory and served as a warning to others to keep away.

As we approached the herd of elephants, the matriarch became irritated by our constant presence. She turned and began threatening us, with her ears held close to her head, trumpeting loudly and aggressively towards us. Then she charged, sending us running in the opposite direction. She stopped after a short distance and stood there, rumbling her discontent. After tossing a large tussock of grass into the air, she rejoined the herd. We waited a few minutes and then continued. All of a sudden a bush roared at me with incredible force. It literally shook. It was a lion and I was within about 12 feet of it. As I froze, the two rangers instantly cocked their rifles and beckoned to me to retreat. I willingly complied. We took a few paces backwards and then turned round. Facing us was a herd of over

80 buffalo. Incredibly, we were surrounded by dangerous animals — elephants ahead, lions to one side, buffalo on the other and a black rhino somewhere in between.

As the reality of my vulnerability sank in, I couldn't help wondering to myself, what the heck am I doing here? The three of us wisely decided to climb a nearby tree and wait for the danger to pass. We must have looked a sorry sight as we all clambered up the largest whistling thorn tree we could find. From our vantage point we could see that there were in fact four lions and they were stalking the buffalo. Thankfully, as the buffalo moved off, so did the lions. We waited a few more minutes before nervously descending to the ground.

Undeterred, we continued our elephant surveillance, eventually reaching a clearing where we could observe the injured baby. It was indeed carrying its leg slightly off the ground and seemed to have damaged its knee. The elephants in the herd were being very attentive to its needs. At one point they crossed a road and the baby struggled to climb up the verge. Each time it reached the top it slid back down. The herd stopped and waited patiently for the baby to catch up. In the end the mother ambled up to the baby and gently placed her trunk between its legs and beneath its backside. She didn't lift the baby up but just gave it enough support to allow it to scramble over the verge, its reputation intact. We reported our findings to the KWS, and they asked us to keep an eye on the baby and keep them informed. I was pleased to tell them a few weeks later that it had made a full recovery.

My career seemed to be progressing reasonably well, but not long afterwards I received some terrible news which brought me back to earth with a bump. Our vet, John Richardson, had been found dead in his flat in Nairobi. He had committed suicide. This came as a great shock to us all. John was a very likeable person and had become a good friend. It was also a loss to the chimps as he was the only vet in Kenya with any real chimp experience.

Around the same time, Simon Barkas resigned and took up the position of general manager at Ol Jogi, a neighbouring ranch, owned by the multimillionaire art dealer Alex Wildenstein. David Heath was employed to replace Simon as general manager of Ol Pejeta. For a while we didn't have a local vet, but eventually Ol Jogi employed an overseas vet called Dr Tom deMaar, and Simon then kindly allowed us to use Tom as our local primate vet in case of emergency.

12

Games and Creativity

Living with three chimps was like living in a top-security fortress. Every time we opened our front door we had to make sure that none of the three terrors was lurking outside. Each time we went outside we had to lock the door behind us even if it was just for two minutes. It became a challenge for Sophie to get into the house — a challenge she relished. If ever we dropped the door key when Sophie was around, she would make a dash for it, and usually get there first. Then she would hot-foot it to the door and have it opened in less than three seconds — a technique she had perfected to a fine art. She would then go on a mad rampage through the house, clambering over the chairs with her muddy feet. Then she would rush into our bedroom and climb the wardrobe before diving off the top, often performing a somersault in mid-air before landing on our bed, leaving her muddy foot marks all over our blankets. In short, she would try to enjoy herself as much as possible, making the most of her limited time in the house because she knew we would throw her out as soon as we caught up with her.

Sophie didn't appreciate the house being out of bounds, and found having her wings clipped quite frustrating. She would regularly look through the windows and check to see where we were or what we were doing. Because the house was circular — like a large African hut — she could see the front and back doors at the same time. The front door was directly opposite the back door, which led on to our patio. If she looked inside and saw that we'd accidentally left one of the doors open, she would tear around the house and dive through the door. And if there was any food in the house she would almost certainly find it. Audrey often made banana cake — everyone's favourite, including Sophie's. We couldn't keep this in the kitchen, so we would hide it in the house in a drawer. She always knew when this was being made, as she could smell it a mile off. Then she would go out of her way to steal her share from the house.

Over time the doors of the house took a battering from the three chimps. Sophie liked going on regular food raids in the kitchen with her two accomplices close behind. Her favoured method was to break the door open using a flying double kick. More often than not, this would splinter the door frame. Once inside she would rush for the cupboard, steal as much food as she could carry, usually a bunch of bananas, and then skedaddle, with the staff in hot pursuit. One of her regular escape strategies was to jettison half of the food in full flight. So she would break off half the bunch of bananas and throw them over her shoulder. This was a conscious decision on her part and was normally successful. She had worked out that as long as she didn't take all of the food she would probably get away with it. This also gave her valuable time to make her escape, as we would have to pick the food up to prevent Tess or Naika from pinching the rest.

One thing Sophie loved even more than bananas was meat. She adored red meat, especially raw beef, and large quantities of it, cramming as much as she could into her mouth. The way she devoured meat was in direct contrast to how she daintily nibbled her chocolate. We often bought zebra meat for the cats, as it was cheap and readily available. This we boiled and placed in plastic bags before storing it in the fridge. Most Kenyans won't eat zebra, partly for religious reasons, as it's a cloven-hoofed animal, but also because it smells so hideously repulsive when boiled. Sometimes Sophie would go on one of her raids and steal the meat before we had got around to boiling it. We didn't allow her to eat raw meat for fear that it might be off or carrying some disease.

Part of preparing the three chimps for their lives in the sanctuary was familiarising them with certain wild foods. An important source of protein for chimps in the wild is insects such as ants and termites, and Sweetwaters was alive with both. Ants and termites are a good example of convergent evolution, in which unrelated species evolve similar traits in response to similar environmental conditions. Ants and termites are unrelated species, yet both form complex and highly co-operative societies and are convergent in many aspects of their social organisation.

When I offered the chimps termites for the first time, they were horrified, recoiling in revulsion as if to say, 'Yuck! You must be joking. And where are my bananas?' I even forced myself to eat a few raw termites in front of them, hoping this would encourage them to try. No chance! And in case you are wondering, they tasted disgusting — like damp soil. I wanted to spit them out but pretended to enjoy them.

But I must recommend the emerging queens, which are delicious when fried, tasting a little like popcorn, and nutritious, being high in protein. Each year the queens swarm out from their mounds in their millions and are feasted on by many species of animal. I would be among those animals and would rush around the garden collecting them. Then I'd fry them and share them with Audrey, Oliver and our fussy chimps. They may not have been particularly enamoured with eating live insects, but they were quite partial to them fried in garlic.

Like all chimps, the thing Sophie liked doing most was eating. In addition to stealing meat from our kitchen, she also enjoyed the less energetic demands of gathering fruit. I would often find Oliver and the chimps picking berries from the many thorny bushes at the bottom of the garden. But every now and again Sophie would decide to go in search of tastier fruit. After the encounter with the elephant, our garden had recovered and was back to its productive best, so she would go on little excursions to our vegetable patch to see what she could find to eat. One of her favourite pastimes there was picking strawberries. We had a large bed of these, which she would regularly visit, plucking out all the ripe ones. More annoying was her habit of pulling up and eating our carrots. A telltale trail of discarded carrot tops would be left littering the path, and we knew it wasn't done by rabbits.

There was one time when Sophie made the fatal error of sampling a red chilli. Her reaction was immediate, and she set off in a screaming panic, with her tongue hanging out, and spitting furiously. I ran after her with a bottle of soda from the fridge and forced her to take large gulps until she calmed down. Sophie learned from her mistake and never repeated it. Maybe the elephant had had a similar experience on a previous garden raid, and I would have been curious to witness its reaction.

The only other time I had seen Sophie react in such a way was when she inadvertently lay down on the grass across a line of safari ants. Again she ran off shrieking in terror. The most difficult part was getting her to lie still while I went through her hair, removing several dozen biting ants.

That same day the ants marched into a wild bees' nest that was situated in a small hole in our *Maerua* tree. As the ants attacked the hive, attracted by the honey, the bees all swarmed out. This was quite handy as I had wanted to get rid of the bees for some time for fear that Oliver would get stung. They swarmed around the garden and then, Sod's Law, flew into our kitchen, where they set up their nest in one of our cupboards. Now they were even more of a nuisance. Sophie was fascinated by these intruders into her kitchen, and we were at pains to curb her curiosity and prevent her from being stung as she insisted on returning to the cupboard for a closer inspection. That evening Steven helped me carry the cupboard out into the garden. Fortunately, the bees got the hint and flew off into the bush. In the wild, chimps extract honey from bees' nests by using tools. These are normally sticks, which they place into the hive to scoop out the honey before licking it off. African bees have a reputation for being highly aggressive, so this was one skill I wasn't about to teach our chimps. This one they could definitely learn by themselves.

Oliver had his own unpleasant encounter with safari ants. He came screaming into the house one day, with a dozen or so in his hair, their jaws clamped firmly to his scalp. As fast as we could, we rummaged through his hair until we had removed every last one. But Oliver was still crying, saying there was one in his eye. We tried to reassure him that there were no ants on his face, but he insisted that his eye was being bitten. Then we noticed one hanging directly between his eyelashes and biting into his eyelid. We quickly removed it. Then, out of curiosity, I placed one of the large soldier ants on my thumb to see how just strong its bite was. Its powerful jaws were on either side of my thumbnail, and quickly clamped on to my thumb. Within seconds the ant had sliced through my flesh like a mechanical metal cutter. I quickly wrenched the ant from my bleeding thumb, and in that painful moment realised just how formidable a predator they really are.

In the wild, chimps have devised their own way of dealing with biting insects such as ants and termites: they generally eat them — only carefully. Chimpanzees are among the few animals that habitually use tools. They fashion tools from sticks, to fish for underground termites or to extract social ants from nests. The insects attack the invading stick, and once their jaws have clamped on to this, it is removed and the insects are eaten. They have also been observed to use sticks and leaves to extract honey from nests and marrow from bones; 'sponges' made of leaves to soak up fluids; and hammers of wood and stone to crack open nuts. Although chimps will sometimes use sticks and stones as weapons to deter predators or to dominate opponents, most of their forms of tool use serve a similar primary function — to obtain food or water. The reward of successfully obtaining food acts as an incentive for this material culture to be preserved and transferred among members of a community and between neighbouring communities.

In 1994, during my time at Sweetwaters, I completed my MSc in Behavioural Ecology. This meant returning to England for short trips to attend courses at Manchester Metropolitan University, including a one-week field course on the island of Tiree, off the west coast of Scotland, studying greylag geese. My main thesis, however, was conducted on the chimps at Sweetwaters. My studies demonstrated that the orphans had suffered varying degrees of maternal and peer deprivation, and, as a result, had missed out on learning important skills such as building nests, sexual behaviour and social grooming. This loss manifested itself in certain aberrant forms of social behaviour that they displayed as adults. The chimps could be categorised into two distinct groups: those that had been kept in peer groups after they had been confiscated by the authorities, and those that had been kept for many years in complete isolation. The ones kept in isolation had high levels of self-grooming and low levels of social grooming, and tended to be sexually inactive, whereas those kept with other chimps showed relatively normal behavioural traits. Few of the orphan chimps attempted to build nests in the trees. Another peculiarity was that they used tools for grooming.

Several of the chimps broke off sticks from the bushes, and used them to clean cuts, remove scabs, splinters, dead skin and ectoparasites, and to clean their teeth, inside their ears, around their eyes and under their nails. They fashioned their tools to a preferred size, frequently reused them and regularly carried them around for long periods in their mouth or hands. I personally observed seven of the orphan chimps using sticks as grooming tools and five of these used them habitually. I observed one chimp, Safari, on 13 occasions, using tools in social grooming of two other male chimps, Poco and Socrate. It may be that this grooming using tools developed because the chimps had been deprived of social grooming, with the use of these tools perhaps acting as a substitute for this behaviour.

The chimps also used sticks for intentional communication, to signal to others that they wanted to be groomed. Barook, Judy, Poco and Charlie would often try to get the staff to groom them inside the building, by presenting the sticks through the bars and then pushing their body up against the bars. Normally the staff would take the stick and use it to groom the chimp. But if they ignored their requests the chimps would often toss the sticks on to the floor towards the keeper. Judy would also use tools to communicate with other chimps, usually when she'd been unsuccessful at soliciting grooming from them. I often observed her requesting grooming from Barook by following him around while pinching a part of her body and holding it towards him. If he ignored her, she would repeat the gesture but amplify the signal by using a tool to groom that area of her body while leaning that part of her towards him.

Barook's use of tools to communicate was even more impressive. On four separate occasions when I was watching the chimps, I noticed Barook sitting on the grass staring at me, some 40 feet away. After getting my attention he walked towards a bush and broke off a branch about three feet long. He then bit off the ends and pulled the smaller branches and leaves off until he was left with a short stick. He then returned to where he was previously sitting and held the stick out to me at arm's length for about 20 seconds while staring intently at me. As soon as he knew he had caught my attention, he walked, glancing over his shoulder at me, inside the chimp building and waited for me by the bars, grooming himself with the tool. On my arrival, he immediately tossed the stick towards me and showed me the part of his body he wanted groomed by pressing it against the bars towards me — be it his head, chest or back of the shoulder. Barook's intentional use of tools to direct my attention to his desire to be groomed was a preverbal effort at communicating with me — a process of communication defined as 'proto-imperative'. In all these incidents the tools appeared to symbolise the act of grooming for the chimps.

Barook's thought process was complex and could clearly be broken down into several components. He saw a human and wanted that human to groom him. Being incapable of speech, he needed to find a way of communicating his intent, and so went to find a sign that the human would recognise. After shaping the sign into something symbolising his desire for grooming — a grooming tool — he showed this to the human. Realising that the human had seen the sign, he went to the only place where the human could groom him — inside the building. Before entering, he gave the human an encouraging 'follow me' glance. Then he entered and waited for the human to join him. Whereupon he gave the grooming tool to the human and pointed to where he wanted to be groomed. Now can anyone tell me that chimps don't possess the capacity for language?

Language and culture are phenomena usually

attributed solely to humans, and often used as defining characteristics for distancing ourselves from other animals. But, as we have seen, apes are capable of communicating using quite complex language, albeit non-verbal, and there is much cultural variation among chimpanzee communities in the type of tools and the manner of their use. And if culture is defined as a trait that is transmitted throughout a population through social or observational learning, then chimps have culture too.

It is true that human culture varies enormously in such patterns as language, diet, dress, art and numerous other social traits. And it is beyond question that humans express culture more than any other species, and that this, coupled with our ability for language, sets us apart from all other species. But, equally, it is wrong to suggest that all traces of culture and language are unique to humans.

After humans, chimpanzees demonstrate the most cultural variation, albeit on a much more rudimentary level. There have been documented as many as 39 different behaviours that vary culturally across different chimpanzee communities. These include behaviours such as tool use, feeding, grooming and manners of courtship that are expressed in some communities but not in others. In the Tai Forest, in the Ivory Coast, chimps use stone tools to break open certain hard-shelled nuts from species such as *Coula edulis* and *Panda oleosa*. They use a hammer-and-anvil technique: the hammer usually consists of a small rock or piece of wood, and the anvil is a larger hard surface, such as a rock or the root of a tree. The nut is placed in a depression in the anvil and hit with the hammer, until the outer shell is broken. This skill is learned by the infants through direct observation of the adults. At Gombe, despite these tools being freely available, the chimps never use tools to open nuts. This stone-tool behaviour is found only in the western chimpanzee subspecies *Pan troglodytes verus*. However, it is not found throughout the *P. t. verus* range, and ceases at the Sassandra-N'zo River, in the Ivory Coast, even though both the tools and the nuts are found on either side of the river. This demonstrates that this behaviour is culturally rather than genetically transmitted.

⋆ ⋆ ⋆

Visitors were always enthusiastically welcomed to our house by the three chimps. Sophie was especially pleased to see new faces. Apart from all the fuss and attention she would receive, she knew that visitors meant tea and biscuits. As soon as the guests had settled down to our tea or soda drinks she would run to the kitchen, grab herself a cup, return and beg from each of the guests in turn, with her cup held outstretched in her hand. The guests would feel obliged to share their drink with her. Her preferred drink was soda. She would walk off with her bottle and slowly drink it, savouring every drop. When it was finished, she would look into the top of the bottle in case she had missed a drop somewhere.

It was interesting that she never thought of looking through the bottle itself. It was if she couldn't grasp the concept that the glass was transparent. On a couple of occasions I saw her accidentally drop her full bottle and spill some of her drink on the grass. Both times she whimpered loudly to herself in despair as she desperately tried to drink the juice from the ground before it disappeared. You could almost hear her thinking, Oh no! She was unaware that I was watching her, so her whimpers weren't directed at me.

* * *

But she didn't just wait for guests to arrive in order to get her favourite tipple. If no drink was forthcoming, she would take the initiative herself. One afternoon she shocked the staff by taking tea from the cupboard and pouring some leaves into the kettle. She then held the kettle under the tap, opened the tap and added water to it. This is how the Kenyans tend to make tea, using hot milk, water, tea and often a touch of ginger, all boiled together in one pot. Thankfully, Sophie never learned how to light the gas cooker to boil the water.

One of our visitors was Jane Goodall. She had come to see how things were progressing with the building of the sanctuary. She visited our house and met Sophie for the first time since she had seen her at Chester as a baby in my arms.

When Jane mentioned that she was going to have an exhibition of paintings by chimps and auction them to raise money, I told her that Sophie enjoyed putting brush to paper. Jane wanted to see if we could get her to do a painting, to add to her collection. So we assembled some of Audrey's watercolours and some art paper and placed them in front of a puzzled Sophie, who, believing we had just given her some multicoloured sweets, gratefully set about eating the paints. I tried to show her what we wanted her to do, by putting the brush in her hand and making her dip it in the paint and then on to the canvas. But Sophie was always suspicious of anyone manipulating her into doing anything she wasn't prepared to do willingly, and instantly pulled her hand away. This had always been her reaction, apart from the time with the baby vervet monkey. Instead she had her own ideas on painting — why throw away good paint when you can eat it? So she went back to what she did best — cramming the paint into her mouth as fast as she could. We quickly gave up on the idea as her creative side obviously lay more in her stomach.

⋆ ⋆ ⋆

Apart from her family, the house staff and the chimp keepers Dickson and David, Sophie had her own few select friends. Although I knew that she had an excellent memory, you could always tell her personal favourites by how she reacted when greeting them. It was usually with a tight hug and an open-mouthed kiss on the neck or shoulder — something she would never do to people she had met for the first time. Then she would normally take them by the hand and drag them to the back of the house. It was always flattering and moving for those chosen ones to be remembered and appreciated. Particular favourites were Tim Hoolahan and his partner, Denise Phelps, close friends of ours, who had spent a week with us when Sophie was a small baby, and another week a year later. Sophie instantly recognised them and was happy to see them again in her new home. Kath Edwards, who was studying for a PhD at the reserve, was another she was particularly fond of. Each year Kath would come to Sweetwaters for a few months to collect her data and then return to England. And finally there was Barry Stevens-Wood, my lecturer at Manchester Metropolitan University, where I gained my MSc.

Barry arrived in Kenya in 1995 and stayed for a few weeks with us at the reserve to oversee some of his university students who were studying the wildlife there. I picked him up at Nairobi airport and on the drive up to Nanyuki told him that Sophie would be very pleased to see him again, as it had been over a year since his last visit. However, as we arrived at the house we met several visitors in the garden. Audrey was entertaining the adults and Sophie the children. After saying hello to Audrey and Oliver, Barry was half-expecting Sophie to rush up to him. I couldn't help but notice the look of disappointment on his face as she took one brief glance at him and then continued to play with the children.

In a way I was more surprised than Barry, as I knew Sophie had an excellent memory and never forgot a face. In fact I believe chimpanzees are better adapted for remembering individuals than humans. They have the added advantage of using all their senses: they recall individuals by appearance, sound and smell. We tend not to use smell except in extreme cases. But a few moments later Sophie turned her head and looked at Barry again, looked away, then looked at him again a split second later and, realising who he was, leapt into his arms. She then gave him a greeting pant and a very affectionate kiss on the neck. Barry was overjoyed, but it was obvious that Sophie must have had a delayed reaction and thought to herself, Wait a minute, isn't that my long-lost friend with the beard? Watching her do a double take was fascinating, as her body language seemed so human.

Yet Sophie never took to my boss, Russell Clarke. She had met him on many occasions, and whenever he and his family visited the reserve they would always pop over to our house for a drink and a chat, and, of course, to see Sophie. Audrey and I both got on very well with them. Russell, an imposing Scot, well over six foot tall, made a lot of effort trying to make friends with Sophie, but always to no avail. She treated him similarly to the way she treated my mother. I am not sure if it was because he was so tall or because of my body language in the presence of my boss, but she never accepted him as a friend. The chimps at Chester had a similar reaction to the zoo's director, Dr Michael

Bramble. If they saw him approaching their island, they would immediately get agitated and start looking for unpleasant things to throw at him. The chimps seemed to view Dr Bramble as a threat to their own dominant hierarchy, yet they had no way of knowing who he was in terms of the zoo hierarchy. He was just a man in a suit whom they saw on a regular basis. The only thing they could observe was the staff's body language towards him.

Even though Sophie enjoyed the company of visitors, she was still very jealous of anyone touching me. She had to know you first, and more or less give her approval. If she didn't know someone and saw them put their arm on my body, she would let them know that they had overstepped the mark and protest angrily. If someone wanted to say hello or goodbye with a handshake, she would pull our hands apart. Being both inventive and mischievous, she soon realised that it was actually quite fun telling people off. As she grew older she improvised, drawing on her rediscovered confidence and status, and invented a game which we named 'the jealousy game'.

This involved manoeuvring people into a position where she could tell them off. She achieved this by pulling people towards me and then, when we were close enough, forcibly pushing them on to me so that our bodies bumped into each other. She would then immediately jump in, angrily howling her displeasure, while trying to grab hold of the guilty party. If I were carrying her she would lean her body out towards her target so that her weight pulled me over to them, as if she were steering a yacht. When we were close enough she would grab the other person's arm and fling it on to me; or, better still, if the person was a close friend, put her hand around his or her head and with her other hand around mine, pull our faces together to make us kiss. She would then give off her loudest scream of disapproval and try to hit them on the face. She played this game so often that I would normally warn my friends in advance. But she had no problems with me holding or cuddling Oliver, although she was a little jealous if she saw me showing any affection towards Audrey and would try gently but subtly to pull us apart. This conduct is consistent with the kind of parent-offspring conflict seen in the wild, where infant chimps often try to interfere with males trying to mate with their mother.

As well as being jealous, Sophie was also touchingly protective of me. If ever she saw me fall over she would scream and run as fast as she could, whimpering, to throw herself on top of me, clutching me as tightly as she could while staring into my eyes for reassurance that I was OK.

Some of Sophie's jealous, protective nature seemed to rub off on Oliver. He had a small group of friends who often came round to the house to play, and one of these was Simon Kenyan, a boy much bigger and stronger than him. Simon was playing with Audrey, when he accidentally pushed her over. There was no malice intended and Audrey wasn't hurt. In fact she found it quite amusing. But Oliver wasn't amused and flew into a rage. He ran up to Simon, pushing him and throwing punches, and yelling, 'How dare you hurt my mum!'

Oliver had to be restrained from his act of gallantry in case he hurt young Simon. He was never an aggressive boy and this was the only time we ever saw him behave in this way. Audrey was surprised and secretly touched by his reaction.

Like most children, Sophie was very creative in her play. Another game she invented was leapfrog, where she would grab Oliver and forcibly bend him over and then quickly run behind him and leapfrog over him, sometimes doing a somersault before hitting the ground. Oliver was very tolerant of Sophie and her annoying games, and didn't seem to mind being used as a vaulting horse. Luckily, she was always gentle and careful not to hurt him — quite unlike Naika, who, when she got excited, could sometimes be too rough and clumsy with Oliver, pulling him around until he fell over and cried. Sophie didn't like to see her young friend upset and would become distressed herself.

But it wasn't just with humans that she had interesting relationships. Her attitude towards the car was intriguing. She loved nothing more than to go for a drive. I only needed to say, 'Come on, Sophie, let's go' and she would rush to the car, open the door, get in and close the door behind her. On the few occasions when I had left the key in the ignition, she managed to start the car, but, thankfully, stalled it each time, as I normally left it parked in gear. During the drive she would sit quietly, concentrating and relishing every minute as she watched the animals and the scenery through the window. Strangely enough, Sophie was also very protective of my car and wouldn't let anyone touch it, threatening to bite them if they continued after she had warned them. If ever the car needed a push-start or it had got stuck in the mud (a regular event during the rainy season), she would even try to stop me from pushing it by pulling my arms away. Whether she saw the car as a living animal and part of the family, I can't say. Only she knew the answer.

She was enthralled by anything to do with the car. During the rainy season, nails or bits of wire previously buried beneath the sun-baked ground rose to the surface. These would often puncture my tyres. On one wet muddy day Sophie observed Steven changing one of the tyres, and was fascinated by what he was doing. She saw him perform this exercise only once, but as soon as he had finished, she grabbed the wheel brace, put it correctly into each of the wheel nuts and tried to undo each one in turn. That was as far as she got, as fortunately they'd been put on too tightly and she couldn't get the correct leverage to loosen them. But it showed that she was a quick learner.

Sophie also greatly enjoyed going for bike rides with the chimp staff. She would sit on the saddle and they would cycle up and down the road with her. She normally initiated these rides by grabbing one of the staff by the hand, dragging them over to the bike, picking the bike up herself and, in the same motion, jumping on to the saddle. The bike would fall over and she would repeat this sequence until the staff member finally gave in to her demands.

Oliver also wanted to ride a bicycle. When he was four years old we bought him a child's bike for his birthday, and somehow managed to wrap it up in paper. It was pretty obvious what was underneath, but Oliver acted as if he didn't know. He loved his present and, like a typical dad, I spent many hours teaching him to ride. We found the best place was out in the reserve, on a long flat stretch of road. The three chimps would come with us in the car, sitting in the back. Then we would let Oliver ride his bike in front of the car. He would pedal along with a big smile on his face for a mile or more, while we followed behind, looking out for lions.

* * *

Anyone visiting our house for the first time by day would probably fail to notice any special relationship between Sophie and I, as she was quite independent of me during the hours of daylight. But in the evening they would be left in no doubt, for then there was only one person who existed in her life. That was the only time when she really needed me. We were living right on the equator and so night fell quickly, at about seven o'clock. Every day at around six she would come looking for me. She would start by standing on tiptoe and peering in through the window to see where I was, groaning as she did so. If I didn't immediately react to this, her groans would become louder until she ended up performing her 'dying swan' act, moaning and groaning and rolling over on the grass below the window. She would continue this until I gave in and came outside.

Normally I would carry her around on my back or hold her to one side. She would wrap her legs around the side of my waist and I would support her with an arm around her back. If I took my arm away, she would grab hold of it and place it back, so that I continued to support her. I then had to stay with her each evening for at least an hour before she would allow me to leave her in her room.

Ever since she'd been a baby, putting Sophie to bed had always been a long, gruelling exercise combining patience and assertiveness. Sophie hated being tricked into going to bed, and would become very distressed. In the wild, a mother sleeps with her infant until it is about five years old, so it was not surprising that Sophie was no different. I had tried for many years to wean her off me at night, but with little success. Even though Tess and Naika were in the room with her, she still wanted to sleep with me. From her point of view, it increased her insecurity if I ran out of the room, as she had no way of gauging whether I would come back. However, she felt confident that if I purposely left her somewhere, then I would come back for her later. This was the procedure I always adopted. I had to get her acceptance that it was time to go to sleep, then eventually she would give in, relax her powerful grip on me, begrudgingly take her blanket and stay in her room while I closed and locked her door.

Sophie was never fully at ease during the night, and if she heard strange noises from outside, such as lions or hyenas, she would become quite alarmed and distressed. Often I would hear her rocking in the night if a noise had disturbed her. One night I was woken by the three chimps all squealing. I grabbed my torch and ran outside to their sleeping quarters, to find them hanging off the rafters. Shining my torch around, I saw that there were thousands of safari ants all over the floor. I carried the three of them into the house and let them sleep on a mattress in our spare bathroom.

In the mornings Steven would let them out of their room early, at around seven. It could be quite chilly at this time of day, so they would all look forward to their mugs of hot tea or cocoa. Naika was especially impatient and would scream endlessly until she had her mug in her hand. If it was a little too hot, then Sophie had a strange habit of taking tiny sips and covering her eyes with her free hand. Then it was time for their breakfast consisting of fruit and *ugali*.

The French have a saying: 'The British eat to live, but the French live to eat.' Well, chimps have a Gallic outlook on life, and there is nothing that gives them more pleasure than eating. Other than breakfast, they received two main meals a day, normally a mixture of fresh fruit and vegetables, such as bananas, mangoes, oranges, passion fruit, papaya, maize on the cob, carrots, onions, ground nuts and sugar cane.

The three chimps differed greatly in their personalities. Tess was the quiet, affectionate one who enjoyed her own company; Naika the loud, abrasive one; and Sophie the smart, cheeky one. Sophie was forever leading Oliver into mischief. One day, when I was in the garden trying to locate the distinctive call of a brown parrot, I noticed the two of them lying flat on their stomachs on the patio floor. Something had caught their fascination, for they were staring intently. Their motionless bodies soon attracted my curiosity. So I went over to see what could be so interesting.

As I approached I could see their noses almost touching a black stick. But as I got closer the stick soon took the distinct form of a black snake, about three feet long. I grabbed both of them and moved them out of harm's way. I explained to Oliver how dangerous snakes can be and then gave out several warning coughs towards the snake for Sophie. Oliver told me that Sophie had poked the snake on the top of its head, and then leapt back when it hissed at her. I picked the snake up with a long stick and carried it out of our garden and into the bush.

Even more alarming was when Sophie decided to lead Tess, Naika and Oliver on an excursion. After a short while we realised that it was very quiet in the garden, so we called out to Oliver, but there was no response. We asked our staff if they had seen any of them. The worried looks on their faces told us the answer and we went on a frantic search outside the perimeter of our compound. We were all calling out Oliver's name but there was no reply. I was really frightened for him. I had seen a large troop of baboons drinking by the dam only a short while before. Baboons have been known to attack and kill children and the thought didn't bear contemplating.

To our immense relief, Steven eventually found Oliver and the chimps not far from the dam in one of Sophie's preferred hiding places. They were climbing to a height of about 12 feet in one of the taller bushes. As this area was out of sight of our garden, it was also favoured by the many lions, elephants, rhinos and buffalo that visited our dam every day. Oliver got a big cuddle from a relieved Audrey, followed by a ticking off, and was told not to put so much trust in that naughty Sophie.

But Sophie's influence was not only bad. She could also be very caring and vigilant. Oliver spent a lot of time climbing trees with his chimp friends. He became very athletic and could climb quite high for his age. Sophie was very attentive and protective towards her younger sibling Oliver and would be concerned for his safety. When they were climbing trees together she would bend branches down for him to grab hold of. Then she would climb closely behind him in case he got into difficulty.

Our garden had several acacia trees, and consequently there were many thorns hidden in the grass. Despite the many warnings we gave him, Oliver always insisted on running around outside barefoot. Not surprisingly he often hobbled sobbing into the house with a splinter. Sophie always became disturbed if she saw Oliver upset. But on one occasion she came to his rescue by successfully removing the thorn herself, using a whistling acacia thorn. She was also seen using one of these to remove splinters from her own feet. You could always tell when Sophie was concentrating: she would sit quietly on the grass with her tongue sticking out, patiently working away at whatever she was trying to do. She had watched Audrey using needles to remove thorns from her own feet. However, she improvised and adapted this observation without any prompting from me or any other person. Sophie also regularly used sticks as toothpicks, especially after eating mangoes. The sticks she used were usually short pieces of *Euclea divinorum* picked up from the ground. Interestingly, the local Kenyans use these as toothbrushes.

Sophie became very adept at using tools, even though I never tried to influence her or teach her any tricks. She often amused herself by using a hammer to bang nails into the hard ground or a piece of wood and then using the other end to claw out the nails. At other times she would go into the kitchen and help herself to a sharp knife, walk off with it and choose a branch from our store of firewood. Then she would spend up to an hour or more carefully slicing off slivers of bark and eating it. She never saw any of us doing this and had improvised this technique on her own — more to amuse herself than anything else. When we heard somebody cutting firewood outside the kitchen, we usually presumed it was either Steven or Paul. But often it turned out to be Sophie having fun on her own, using a machete to try to cut a large branch in two. She would sit there for ages with her tongue sticking out and her face a picture of concentration.

Just when all the chimps seemed to be doing well, we hit a crisis. There was an outbreak of shigella — a severe and highly infectious form of dysentery — in the adult group of chimpanzees. Three chimps, Barook, Charlie and Jojo, became very ill. Their faeces were thick with blood and mucous and they had a high temperature with fever. Barook contracted the disease first. We took a sample and sent it to the laboratory in Nanyuki. However, nothing significant was found. The following morning Jojo was ill with similar symptoms. We took another sample and sent it away again, this time for the lab to grow a culture; it would be a couple of days before the results were available. By the following day Charlie was also ill. The next day the lab confirmed that the illness was shigella, and prescribed the drug ciproxine.

That same day I called Charlie's owner, Karl Ammann, to inform him that Charlie was ill. Karl had visited Charlie only a few days earlier when he was healthy, so he was shocked to find him suddenly so unwell. Tragically, Charlie went downhill dramatically and died the following afternoon. It happened so fast and it was a tremendous blow to us all. Karl loved Charlie, and was devastated. We managed to contain the spread of the disease by disinfecting the entire building every day and installing foot baths containing disinfectant for the staff to use when entering and leaving any of the rooms. Barook and Jojo, who had contracted the illness before Charlie, were both still seriously ill. But Barook was as tough as nails and, despite his high fever, continued to eat and drink as normal.

Jojo, however, was becoming desperately weak and refused food or even liquids. She was seriously dehydrated and spent an entire day lying down. I feared the worst and felt sure that she would be dead the next day. The vet, Tom deMaar, decided that she needed to be put on an intravenous drip, although he was a little apprehensive that the anaesthetic might kill her, as she was so weak. But we had no choice, so he administered a low dose of anaesthetic, which gave us just enough time to give her a full litre of liquid. This seemed to have an immediate effect on her, and by the following morning her condition had greatly improved and she had regained her appetite.

Part Four

Becoming Chimp

13

Integration

Sophie, Tess and Naika were now spending most of their days in the sanctuary. They felt quite at home here and enjoyed performing their acrobatics: climbing the tall acacia trees, jumping from branch to branch and swinging on the thick lianas which hung from the acacias like twisted old rope.

I tried as much as possible to encourage our three chimps to forage for wild foods in preparation for their life in the sanctuary. In the wild, fruit makes up a large part of a chimp's diet. Many of the local wild bushes in the reserve had small edible fruit, usually protected by sharp thorns. These berries are very important to the animals and birds in the reserve because they fruit during the dry season, when there is little food available. The biggest and tastiest were the cherry-sized *Carissa edulis* berries, and Oliver and the chimps loved carefully picking them.

In the last week of January 1996 the remaining ten young chimpanzees arrived from Burundi. Like the adults before, they were all refugees — innocent victims of the racial strife that was afflicting Burundi. These younger chimps were already integrated into a group. The oldest male, Max, was aggressive towards the staff and prevented us from entering the enclosure with him. So we decided to integrate him into the adult group and reunite him with his best friend, Alley, with whom he had grown up since infancy. We now had an adult group of 13 chimps on one side of the river and an infant group of nine on the other.

Among the infant chimps were four males and five females. In descending order of age they were: Akela, a female and the oldest of the group; Amizero, a large female a couple of years older than Sophie and the dominant chimp in the group; Uruhara, Toto and Niyonkuru, all males about Sophie's age; two females, Dufat-anye and Chipie, both around the same age as Tess and Naika; a young two-year-old female called Bahati; and the youngest member of the group, a small, 18-month-old male called Kizazose.

Rather than introduce Sophie, Tess and Naika to the Burundi chimps, I decided instead to introduce the Burundi chimps to them. In other words, build the new group around the three. All the introductions took place outside in the sanctuary, with me and the staff sitting among them. Knowing we were there by their side gave the chimps a lot of confidence. For Sophie, it was also in stark contrast to her experience at Chester.

On the last day of January, Audrey, Oliver, Steven and I drove our threesome to the sanctuary. We carried them inside and settled down near a bush which was covered in small black edible berries. Then Dickson and David joined us with the three youngest chimps: Chipie, Bahati and Kizazose. Because Kiza was the baby in the group, he was very popular with all the group members and was constantly spoilt for attention. They were already used to socialising with the other Burundi chimps, so meeting three new faces was no big deal for them.

After a few reticent minutes, the six were soon making their individual acquaintances. Bahati and Kiza were both playful and friendly. Chipie, the least social of all the Burundi chimps, kept her distance and climbed to the top of a small tree and stayed there. The other chimps were more sociable and it wasn't long before they were slapping each other hard on the back and play-chasing. Sophie, Tess and Naika took an instant liking to young Kiza, who relished the attention he was receiving. Sophie in particular wanted to cuddle him, and whimpered whenever he moved away from her. He was enjoying himself and wanted to ride on her back. But Sophie couldn't comprehend what he was up to. Instead she found him ticklish and annoying and would end up laughing loudly before wriggling herself free. In fact she never took to any chimp riding on her back.

Tess, though, was more than willing to carry him. After all, he was a lot lighter than the other male she was used to carrying — Oliver. We kept the six together for a few hours and then separated them. We then drove our exhausted three babies back home and gave them an early night.

The next morning I could see that Sophie was very keen to get in the car. It was clear that she was eager to go and meet her new friends again at the sanctuary. She was looking intently into my eyes for a signal that she could accompany me. I didn't disappoint her. Without uttering a word, I simply nodded my head in the direction of the car. Sophie rushed to the car, the others quickly following. She opened the door and scrambled in and then shut the door behind her, leaving her two disappointed friends outside. I opened the door for them to get in and we drove off.

The six chimps carried on where they had left off the previous day. We kept them together for a week until we were sure they had formed friendships before attempting any further introductions. The other chimps were able to observe all the interactions unfolding and could see that these outsiders were friendly, and this in turn helped to relieve some of the tension during subsequent introductions. For the entire week they played vigorously and non-stop. There was much excitement during the day and tiredness in the evening — so much so that by the end of the week Sophie and Kiza had both come down with colds.

The next chimp to be introduced was the female Dufatanye. About a year younger than Sophie, Dufa was a tough character and could be a real handful. She was also an excellent pickpocket. In fact if I could have got a shilling for every pen or pair of sunglasses she had pinched from visitors, I would be a rich man.

Dufa liked to play roughly and spent most of the day wrestling with her new companions. Niyonkuru was the first male to be introduced. He was a very good-looking individual. Speaking strictly as a jealous parent, if I could have chosen who I liked to father Sophie's babies, then it would have been Niyon. Sophie, Tess and Naika were a little apprehensive of this big male. But he was intent on being friendly and seemed to take a shine to Sophie.

After Niyon, we introduced another male, Toto, again with few problems. Our presence was comforting not only for the chimps but also for me. If things got a bit rough with the males we could intervene or distract them. We kept this group together for about ten days before introducing Amizero in early March. Sophie and Tess were quite nervous of this huge female. Amie was bigger than any of the males and feared no one. She also had an off-putting habit of staring straight into your face from very close up. But by this time most of the Burundi chimps were already in with Sophie, so the remaining few were more interested in simply rejoining their friends.

Uruhara was the last male to be introduced because he was the most boisterous and a bit of a bully. But we needn't have feared as he turned out to be very gentle, cuddling each of them almost immediately. Finally, on 18 March, we introduced the last and the oldest chimp, Akela. She too was very friendly, and before the morning was over was grooming both Sophie and Tess.

Generally, all the introductions went very smoothly. This further helped Sophie regain some of her former confidence. Chimps aren't so bad after all, she must have thought. Naika probably integrated the most easily into the group. In fact she looked as if she had always been with them. Tess, however, remained aloof and peripheral — somewhat anti-social — but maintained her close friendships with Sophie, Naika and the two youngsters, Bahati and Kiza.

What a dramatic change to Sophie's life! She was now in a large group and would have to adapt and deal with all the social dynamics that this entailed. But first she had to come to terms with the fact that she was no longer the boss. This she coped with well, soon learning to be submissive at the appropriate time whenever things became too rough. In general, she was very popular with her new friends, especially the three adolescent males.

Each morning we would take the 12 chimps down to the river, where the tall acacia trees gave them plenty of shade. They loved climbing as high in these trees as they could. They seemed fearless, swinging and jumping from branch to branch, leaving me constantly fearing the worst. Sophie would be quite independent of me, only occasionally checking on my whereabouts.

Whenever we moved off from the river back towards the building, the chimps would normally walk in single file along one of the many footpaths in the sanctuary. Sophie would invariably wait and make sure that everyone had joined the line before taking up the rear position.

Interestingly, Sophie's grandmother, Meg, used to do the same with the chimpanzees at Chester. But whenever we were walking back to the building together, Sophie would always insist that I give her a piggy-back, not just because she was very lazy and spoilt, but also because she was nervous of any impending dangers as we walked through the thick bush. The other chimps would follow quietly in single file. Unlike us, the chimps weren't to know that their sanctuary was relatively safe from predators. The riverine area was more open than the thick bush, where visibility was down to just a few yards. The walk back to the building was more than a mile along a winding trail, in the baking hot sun, which nobody particularly enjoyed — least of all me, with a 66-pound (30 kilogram) chimp on my back.

* * *

Although the fence was effective in keeping the chimps in, it wasn't foolproof in keeping other animals out. With there being so few grazing animals inside, the sanctuary was fast becoming an oasis. It wasn't long before three hippos took up residence in the river. Hippos are among the most dangerous animals in Africa, especially if you come between them and the water when they are grazing on land. The chimps had a lucky escape when they disturbed a hippo as it was grazing on the lush vegetation. The startled animal charged through the centre of the group, missing Tess by just a couple of feet, before diving into the water. After that incident Tess was always nervous walking through the bush and forever on the lookout for danger. If she heard a strange noise she would refuse to continue until the keepers had checked the bushes to see if it was safe.

Sophie was still scared of animals calling in the night. One evening a lion was heard roaring not far from the sanctuary. The following morning Sophie refused to come out of the building with the rest of the chimps until late in the afternoon. On another occasion a group of lions climbed over the fence and into the sanctuary. They stayed inside for over a week, feeding off the small population of warthogs and bushbuck. We had to keep the chimps in the building during this time. After a week the lions were showing no signs of leaving, so we rustled together around 30 people and walked through the sanctuary in a line, making as much noise as we could. We reached the other side of the fence without seeing any lions, and so presumed we had succeeded in scaring them out. But for a while afterwards, travelling down to the river was a tense affair, as we could never be too sure that the lions hadn't climbed back in again.

Some bushbuck had escaped the lions, and one in particular had an unusual interaction with Sophie's group. The chimps were in the middle of their enclosure when a female bushbuck appeared from some bushes and walked straight past the two keepers and towards the chimps. Not knowing quite what to make of this animal, the chimps all climbed into the trees to get away.

After a few minutes Amie decided she had had enough and came down to the ground. With the other chimps watching from the safety of the trees, she chased the bushbuck away. But barely ten minutes later the interloper reappeared, bolder than ever, and started approaching them closer than before. She stood just six feet away from the astonished staff, obviously reassured by the chimps' relationship with them. This time the entire group chased her deep into the bush. She escaped unharmed and didn't return, and we were left to speculate what would have happened had the chimps captured her.

Bushbuck regularly travel on the ground with troops of baboons. They gain security by being with the baboons, who constantly warn their troop members of impending danger by giving off a loud warning bark, similar to a dog's. In fact baboons and bushbuck are said to have similar-sounding warning barks. The bushbuck also benefit from scavenging the fallen fruit, leaves and seed pods that the baboons drop or discard when feeding from a tree. In return the baboons benefit from the bushbuck being an extra pair of eyes. The relationship is made more remarkable by the fact that the baboons occasionally kill and eat baby bushbuck if they come across them. The female bushbuck in the sanctuary must have confused the chimps for baboons.

For many animals, meat is an important source of protein, fat and calories. In the wild, chimpanzees regularly hunt small monkeys, young bush pigs and antelopes such as duiker and bushbuck. They are known to prey on at least 19 species of small mammal. Of the higher primates, only humans and chimpanzees hunt for and eat meat on a regular basis.

Although chimpanzees sometimes hunt on their own, they tend to form large hunting parties, consisting mainly of adult males, although females occasionally hunt. The size of the hunting party can vary greatly from a single individual to as many as 35. The hunts tend to be carried out in silence, and usually involve much social co-operation. If the hunters are successful, their silence is typically broken by an explosion of screams and frenzied excitement.

Other than for the obvious nutritional purposes, chimpanzees also hunt meat for social reasons. Meat is shared among individuals according to the social dynamics of the group. Sometimes a male won't share meat with a particular female until after he has mated her. Interestingly, the number of males participating in a hunt tends to increase in proportion to the number of sexually receptive females in the community. So males appear to use meat sharing as a tool for gaining access to reproductive females. They also use it as a political tool to reinforce relationships. At the Mahale Mountains, in Tanzania, males normally share meat with their allies but rarely with their rivals.

In addition to the protein they obtain from eating small mammals, chimpanzees in the wild also get it from insects, such as ants and termites. At Gombe the chimps were seen to eat insects virtually every day. Both sexes eat insects, but females do so much more frequently, and for longer duration than males. The females seem to concentrate more on feeding on insects, whereas males concentrate on obtaining the bulk of their protein by hunting small mammals. This difference in feeding behaviour between the sexes is vaguely similar to that of certain human hunter-gatherer societies in which there is a division of labour, with the men hunting and the women gathering fruit and seeds and so on.

So I suppose it was hardly surprising that the young chimps at Sweetwaters were beginning to show predatory signs. Baboons were never tolerated in their enclosure and would be pursued by the whole group, usually with Niyon leading the chase. There was a second interaction with a bushbuck, this time involving Sophie and Naika. Once again a young female began approaching them. This time their reaction was immediate, with Naika leading the chase and Sophie directly behind. The two of them chased the bushbuck for several minutes before giving up and leaving her unharmed.

Not long after, the same two were involved in another hunting session. While they were down by the river, a coypu suddenly emerged from the water and clambered on to the bank. They were instantly fascinated by this strange-looking animal. Sophie hid in the nearest bush, waiting for the coypu to walk past her, and Naika went off to find a weapon to throw. When the coypu was within a short distance, Sophie rushed out from her hiding place and made a grab for its tail. She narrowly missed, and the startled coypu ran further along the bank, whereupon Naika came charging towards the animal and threw a stone at it, so that it fled back into the river. Satisfied that it had been given a thorough lesson, they settled down and begun foraging for edible leaves instead. Chasing my cats had obviously been useful in preparing Sophie for life in the sanctuary. But I'm not sure if Sophie would have eaten the coypu or merely cuddled it.

As well as the large mammals the chimps had to contend with, they also had to avoid venomous snakes and spiders. However, they seemed aware of the potential dangers. This was emphasised when Tess spotted a snake swimming across the river. She gave out a loud warning bark, and the other chimps came to see what was wrong. When they saw the snake they started throwing stones and branches into the water. Naika was especially determined, and threw more stones than the rest put together.

Over at the adult chimp group, Ndaronse was bitten on his neck by a spider. We were quite worried that it might prove fatal. The area around the bite became grotesquely swollen, like a balloon filled with pus. By the evening the skin had turned black and the balloon looked about to burst. And that's exactly what it did: the next morning, much to our relief, we found him with the hairs on his chest all matted and strands of dead skin hanging down his neck. It wasn't a pretty sight, but he made a rapid recovery.

Now that all 12 chimps had been introduced, the next step was to get them used to sleeping overnight at the sanctuary. The infant chimps had their own building, which consisted of five rooms, all interconnecting via sliding doors. I promoted Steven to chimp keeper and he proudly moved with Sophie, Tess and Naika from our house to the sanctuary. This was to maintain continuity for the three chimps, and also because Steven was an excellent keeper who had the respect of all the chimps. We replaced Steven at the house with his brother, Karanja, who was trained as a cook.

In early May 1996 we brought all the chimps inside in the evening and separated Sophie, Tess and Naika from the main group in one of the rooms, and kept Bahati and Kiza with them. While they were eating, I left the sanctuary and drove home, leaving them to sleep there for the first time. This period of integration was more stressful for Sophie. She was used to sleeping with her real family at Spoonbill Dam; this is what she wanted to do and where she wanted to be.

A week later Debby Cox, who was in charge of the Jane Goodall Institute in Bujumbura, where all the chimps had come from, paid us a visit to see how they were getting on. Debby suggested making hammocks out of hessian sacks for them to sleep in. This was a good idea and proved popular with the chimps, and they would each claim their hammock before it was dark. Sophie seemed to prefer sleeping in a hammock, and that evening she picked up Kiza in her arms and carried him into her hammock, where they settled down for the night.

The next day Debby wanted to spend some time with the chimps inside their enclosure. Audrey and Oliver decided to join her as well. This turned out to be a big mistake. Oliver had previously been in with all the chimps individually or in small groups but never with them all together at the same time. Amid the excitement and confusion, Oliver found himself surrounded by curious chimps. He felt nervous and Niyon immediately became aggressive, attacking him and biting him hard on the chest before I could intervene. Oliver screamed in agony. I chased Niyon away, and carried poor Oliver outside. He was terribly upset and in a lot of pain. His wound wasn't bleeding but there was deep bruising, which later left a permanent scar. Niyonkuru went down in my estimation overnight and I no longer wanted him for a chimp-in-law.

Just to complicate matters, on 1 June, only three weeks after the introductions, we were offered another chimpanzee, a female roughly Sophie's age called Naika. Her owner, Harash Devani, a wealthy Nairobi businessman, kept her as one of the family but could see that she was getting too big for the house. He had built a large cage for her in his garden but felt that she would be happier living with other chimps in a more natural environment. We now had two chimps called Naika, which would cause confusion, as all the chimps recognised their own names. So we decided to rename the new chimp Maisha.

I wasn't sure how Oliver would react to a large

chimp so soon after being attacked, but Maisha was friendly and helped restore his confidence and faith in his hairy cousins. She was about five years old and had lived with the Devani family since she was a year old without ever seeing another chimp until now. Although friendly, she was very stubborn. Getting her to go to bed was infinitely worse than putting Sophie to bed; it would take several hours before we could coax her into the room. There was no forcing her to do anything she didn't want, as she wouldn't hesitate to bite to get her own way. Fortunately, she took a liking to Steven, so he had some control over her.

The KWS asked us to keep Maisha a month on her own as a quarantine precaution. A few days after her arrival, their vet came and took a blood sample and inoculated her against tetanus, measles and TB and also gave her polio drops. Shortly afterwards Maisha came down with a bad cold. The stress had probably been a bit too much for her. She stubbornly refused all medicine even when it was disguised in the food. In fact she would be so suspicious that after a while she refused any food or drink, regardless of whether or not there was medicine hidden inside. She'd been with us only a week and now I had to call the vet to give her long-acting antibiotics. She made a recovery, but I could see we were going to have our hands full with Maisha.

It was only a couple of months since we had settled the three chimps into the sanctuary, but I decided to bring them back to the house and introduce them to Maisha there. When I opened the car doors for them, it sparked off a race to see who could get in first, and they knew exactly where they were going. Sophie was especially excited and as soon as she was in the car, she hugged Tess, Naika and then me. As we approached the house she gave out loud hoots of joy. The chimps opened the doors for themselves and jumped out. Audrey and Oliver were pleased to see them, too.

Then Maisha appeared with Steven from the back of the house. Sophie was surprised to see a strange hairy face and ran up to see who it belonged to, and more to the point, what it was doing at her home. Maisha was initially nervous, and not interested in being sociable. But, after a lot of effort, we managed to sit the four of them down on the patio, where they soon began wrestling and play-chasing. The three chimps could easily have been jealous of Maisha, but by now they were so used to seeing new faces that they took it all in their stride. Besides, Sophie was so excited to be home that she didn't seem to care. We kept the four of them together all day and that night they slept together at our house.

A few days later I drove to the sanctuary and picked up Chipie, Bahati and Kiza in my car and, along with David and Dickson, drove them to our house, where we introduced them to Maisha. For a few hours we had seven crazy chimps running around like lunatics, turning the house into a scene from a Keystone Cops film.

We began familiarising Maisha with the sanctuary and also the rest of the Burundi chimps. The remaining introductions all took place in the sanctuary without major incident. As she was quite big and confident, Maisha integrated pretty well. However, as with Tess at first, she remained slightly detached from the others.

Now that the chimps were all fully integrated and living in the sanctuary, we were able to live a near-normal life again. Of course, we missed not having our hairy family around the place, but the luxury of being able to leave the front door of the house open soon made up for that. There was no more locking the door every time we wanted to go outside, no more raids on our kitchen for bananas and, best of all, a regular supply of banana cake.

As the months went by and Sophie realised that she wasn't coming home, she began to resist her enforced separation from her family. She knew we weren't far away and was aware roughly where our house was. After all, she'd been paying attention during all those car rides. On two occasions she managed to escape from the sanctuary by slipping under the electric fence. Both times she tried to find her way back through the bush to our house. This was a worrying time for everybody, as it was very dangerous for her to be outside the sanctuary, especially alone. There were at least 40 lions in the game reserve, and many leopards and cheetahs too. Add the countless baboons, 20 black rhinos, hippos, buffalo and almost 100 elephants, and there was a lot for any small young chimp to contend with.

The first time Sophie escaped she was found not too far from the sanctuary. She walked about a quarter of a mile into one of the staff camps, where about 25 staff members all lived together in a little village within the reserve. Several of the chimp staff also lived here. As Sophie calmly strolled into the camp, all the non-chimp staff scattered for safety. Sophie, not knowing what all the fuss was about, settled down outside Dickson's house and began playing with some flowers in his garden. Dickson invited her inside his house and made her some tea. When she had finished he carried her back to the sanctuary.

The second time she took off she couldn't be found anywhere. All the staff started to search the bush. It turned out that she had crossed a weir on the river and had gone over to the other side, where the adult chimps' enclosure was. Then she had climbed up the steps of a wooden viewing platform that we had built for tourists. There she found a radio belonging to one of the staff. After placing this between her hip and stomach — known as a chimp's 'groin pocket' — she carried on walking. Dickson eventually found her near the marsh, more than halfway to my house. He believes that she'd taken the radio in order to listen out for my voice. He asked Sophie, 'Where are you going? Are you looking for Vince and Oliver?' Apparently, she immediately stood on her heels and began looking all around her. Although she was found only about two miles away, she had travelled through very thick bush. We will never know if she would have successfully found her way home. It was quite sad that Sophie should be driven to such extremes to get back to us.

But eventually Sophie accepted that the sanctuary was now her home. She was now a member of a group of 13 chimps, living in semi-natural surroundings in Africa — a far cry from what could have been her home at the zoo. I often remembered the vow I had whispered to her many years ago at Chester and felt that I had finally achieved some of my ambitions and responsibilities as her parent.

★ ★ ★

The wet season came to an end and slowly the grass returned to its lifeless form. On one hot sunny day, while Elizabeth was hanging the washing on the line, Paul gathered all the ashes from our previous night's fire and threw them into the bush. Then he trundled up the road in search of more firewood. However, unbeknown to him, hidden among the ashes were some hot embers. As soon as they came into contact with the dry grass, they burst into flame. With the wind blowing strongly, the fire spread at a tremendous pace and it wasn't long before it had engulfed an area the size of our garage. Not so big, you might think, but bush fires burn at explosive speed, and it proved difficult to contain. Meanwhile, in the house, Audrey was wondering what all the loud, crackling noise was outside. So she went to investigate, and met Elizabeth rushing towards her.

'Audrey, come quickly. The bush is on fire,' she exclaimed.

Audrey went to see for herself. Sure enough, the fire was now burning fiercely. She shouted for Karanja and Paul to come and help. Quickly, they cut branches from a nearby bush and began beating the flames. Oliver was playing one of his computer games when Karanja told him to come out of the house as it wasn't safe. Oliver wanted to help fight the fire too, but Audrey told him to keep well away. While they were beating away on one side, the flames were spreading on the other, and despite there being four beaters they were unable to control the fire. Soon it had become a hot, raging inferno. Realising they couldn't cope alone, they decided one of them needed to go for help. We had only the one radio, and that was with me.

The fire had now engulfed an area over 100 yards long and 40 yards wide and was heading towards the house. The heat sent hundreds of burning leaves soaring high into the air. It only needed one of these to land on our thatched roof and our house would be burnt to the ground in minutes. Everybody was now panicking. Karanja was shouting, 'Fire! Fire!' as they continued beating rhythmically at the flames. Oliver began crying as it seemed more and more likely that our house would be burnt down. Audrey ran up the drive in the hope of meeting a car on the road.

High up on a hill on the ranch was a lookout post. The two men manning it had seen the fire at the beginning and presumed we were burning rubbish. But as the fire grew larger and larger they put out the alarm over the radio, 'Nine-nine's house is burning!'

I was in the chimp sanctuary at the time and couldn't believe my ears. Petrified that my family might be in danger, I rushed with my staff to the car and drove off with them all crammed into the back. Meanwhile other ranch staff were also heading towards our house from different directions. A squad of British soldiers who were training up on the ranch also heard the alarm over the radio. As I approached the drive to my house, I met Audrey on the road. She had just flagged down a car and was about to be driven to the Lodge to get help. She got in my car and we drove to our house, where we found the exhausted and panicking staff frantically beating out the flames but fighting a losing battle. The flames were now only 15 feet from the side of the house.

It was a tremendous relief for Oliver and the staff to see the cavalry coming to the rescue. The other cars arrived together some minutes later. Now we numbered around 20 people and we formed a line and together began beating out the fire. Slowly, it began to retreat, and after an hour or so we had finally extinguished the last flame. This was a narrow escape and a good lesson for everyone. We were all hot and exhausted and celebrated with bottles of cold beer all round. Afterwards we began wondering if Paul was secretly Masai and not Kikuyu after all. The Masai regularly burn the grasslands to promote new grass for their cattle and to control the tick populations.

But the cycle of nature continued and the rains eventually returned as usual, transforming the large burnt area around our house to a carpet of brilliant green. Within a week many antelopes were attracted to this lush new growth, much to the gastronomic delight of our resident pride of lions.

A few weeks later we had a visit from the Walt Disney film company, who were working on their forthcoming cartoon feature film *Tarzan*. They had a crew of about a dozen people who spent a couple of days at our house in order to draw and take video footage of the three chimps and Oliver. They wanted to use them as models on which to base their drawings for the film. They modelled the young Tarzan on Oliver and Tarzan's young ape friends on Sophie, Tess and Naika. Oliver's name and photo were to appear in the book *The Making of the Film Tarzan*.

We decided to do the filming by the dam at our house. This meant we could bring the three chimps back to our house again for a little reunion. Sophie was very excited about coming home. I pointed to the car and said, 'Let's go, Sophie.' She knew instantly that she was going home, and did cartwheels all the way to the car, opened the door herself and sat in the front. Tess and Naika sat in the back as usual. As we approached our drive Sophie gave out hoots of excitement, leant over to me and, with her open mouth pressed firmly to my neck, panted hot breath on to me for about ten seconds. The film crew spent the next couple of days taking video footage and still photos and making drawings of Oliver and the chimps. They were particularly interested in capturing Oliver wrestling with the chimps by the dam.

There was much demand to film the chimps, and each month we received film companies from all over the world. Sweetwaters also had a regular flow of VIP guests who would stay at the Lodge. The identity of the guests was usually kept a secret. Normally I would be around for these trips, but on one occasion I decided to spend the weekend with the family in Nairobi.

We returned home on Sunday evening and were disappointed to discover that the VIPs visiting that weekend were no less than Mick Jagger, the lead singer of the Rolling Stones, his supermodel wife, Jerry Hall, and their children. It must have been very amusing for the Jaggers because absolutely nobody recognised them during their entire three-day stay. Catherine Heath, wife of David, the general manager of the ranch, picked them up at the ranch's airstrip. Mick Jagger walked over and introduced himself: 'Hi, I'm Mick.' But she didn't know him from Adam. She then drove them down from the ranch to the game reserve, which takes about 25 minutes, and attempted to make polite conversation.

'So, Mick, what do you do for a living?' she asked.

'I'm a percussionist,' he replied. Still none the wiser, she dropped them off at the Lodge and drove back up to the ranch. None of the African hotel staff had ever heard of the Rolling Stones either. The Jaggers apparently visited the chimps on two occasions. I wanted to verify with my staff if Mick and Jerry had been with the children when they visited the chimps or if the children had just gone with their nanny. So I asked Joseph Maiyo. He told me that he wasn't sure, as we had had many guests that day. I explained to him who the Rolling Stones were and how famous they were in the West. Joseph recalled that an English family with children had visited the chimps on their own. He then described the man as thin, with a wide mouth and a very lined face. I couldn't have come up with a more perfect caricature myself.

I'm sure that Mick and Jerry must have felt totally relaxed away from the usual glare of publicity that permanently surrounds them. It was a shame I didn't get the chance to meet them, as I would have invited them to my house to play with the three young chimps — something their children would have loved. The next day one of the research students, studying the chimps, gave Joseph a tape of the Rolling Stones as a gift.

14

Community Life

Any primate sanctuary will have an optimum carrying capacity for the animals living inside. Ideally a balance needs to be found between the number of animals and the size of the sanctuary, taking into account stress levels, sex ratios and the destructive capabilities of the primates towards their habitat, notably the trees. One of the problems with any animal sanctuary is that when the conditions are favourable, the animals tend to multiply.

A sanctuary for any species of ape can become a major financial drain. To be sustainable, it needs to be profitable. So making a profit from conservation, if for the right reasons, is not necessarily a bad thing. Lonrho viewed the chimp sanctuary as a non-profit venture. At the same time they were hoping to benefit indirectly from an increase in the number of tourists visiting the game reserve and from those staying at the Tented Camp.

In order to oversee all the major policy decisions and long-term management of the chimp sanctuary a steering committee was set up. One of the major bones of contention surrounding the sanctuary concerned the definition of the breeding policy. Jane Goodall and

Richard Leakey, both members of the committee, agreed on a policy of limited and controlled breeding of selective females, and I was in favour of this policy. But after Richard Leakey was removed from his position as head of the KWS and replaced by Dr David Western, he resigned from the steering committee. Not long afterwards Jane Goodall also resigned.

The new steering committee was against any form of breeding. They believed that allowing the chimps to breed would be taking up places in the sanctuary that could be given to other, more needy orphans. This is fine in theory. But, in practice, once you have established a settled group of adult chimps, to introduce infants to that group is to put their lives at risk. Anyone knowing or working with captive chimps will also recognise the positive impact that infants have on a community as a whole in terms of reducing stress levels and the overall enrichment they provide. Besides, we weren't advocating uncontrolled breeding — but merely that one or two females in both groups should be allowed to have one infant and that the rest be placed on long-term contraceptives.

* * *

Away from all the politics, the three chimps were settling in well with their group and adapting to their new life in the sanctuary. Sophie had become the most active groomer among the chimps in her group. Apart from providing a valuable service to her partners, she also obtained a small amount of protein every day by eating the many ticks that she found in their hair, often engorged with blood. She would groom little Kiza more than anyone else, and would often carry him off to a quiet corner and groom him there, away from the others. Although Kiza had been seen riding on Sophie's back for a short while, this proved to be a rare event, and she was more likely to carry him awkwardly in her arms. When walking back from the river, Kiza sometimes became tired and flatly refused to take another step. He just stood there whimpering non-stop until someone came to his aid. Tess, Maisha and Niyonkuru were the chimps who would normally carry him the rest of the way. This was no mean task for the young chimps, as it involved walking more than a mile.

Over the months Uruhara had become very attached to Sophie. The two other males, Niyon and Toto, enjoyed tormenting the females. But if they picked on Sophie, Uruhara would normally come to her aid. He would even become upset if he lost sight of her and begin whimpering. One day, after the chimps had spent several hours by the river, it was time for the staff to walk them back to the building for their food and then to bring them inside for the evening. They assembled the chimps together ready to walk them back, and to their surprise Sophie was nowhere to be seen. She must have gone back to the building on her own, they decided. But when they reached the building she wasn't there either. It was gone five in the afternoon, and there was less than two hours of daylight left. So they rushed back down to the river again. Still no sign of her. They began to panic, calling out her name, but nothing stirred.

They returned to the building and continued calling her name. The other chimps had all finished eating and were now settling down to sleep. It was now after six and the sun was setting. All of a sudden Uruhara got up and started walking towards the river on his own. The staff decided to follow him. He walked all the way to the river bank and straight up to a thick bush, where he found Sophie tucked up comfortably and getting ready to sleep. Uruhara had somehow known all along where she was, and had realised that the staff were looking for her.

The chimps enjoyed their time by the river. The loyalty between Sophie, Tess and Naika remained as strong as ever, and if one of them was bullied the other two would come to the rescue. Naika was becoming much calmer as she adapted to life in the sanctuary. But she was still as greedy as ever, eating her food faster than anyone else and then begging more from the others. Sophie and Tess often gave her food, just so they could get some peace from her constant whining. But Naika would never share her own food, and if anyone came near her when she was eating she would scream and gesticulate angrily.

For her size, Naika was well proportioned. She was also strong-willed and wouldn't tolerate rough play from any of the larger males. If things got out of hand she would quickly lose her temper and try to get her own back, even if they were twice her size. The other chimps had a healthy respect for her temper. When she was bitten by Niyon during a squabble she spent the next three days trying to get her revenge. Niyon was aware of this and wisely avoided her at all times. During the heat of the day the chimps would regularly come down to the bank to drink from the river. But in the dry season, when the water level had dropped, the water was often out of reach, so many of the chimps had problems drinking. This was never a problem for Naika, who would dip a stick or a handful of leaves into the river and hold them above her head so that the drips fell into her mouth. She would also reach down and fill her cupped hand and then raise it to her mouth. Dufa noticed her doing this, and came and sat next to her for a closer look. Dufa was so fascinated by this behaviour that when Naika had finished drinking, she wouldn't let her leave and kept grabbing her hand and pushing it back into the water.

Apart from Sophie and Naika, the only other chimp in the infant group to use tools was Niyon. He had an unusual habit of using a stone to pound the hard sugar cane to make it easier to eat. He also devised an unusual way of eating oranges. He would place his orange on the floor and then stamp on it, splitting it open so that he didn't have to peel it.

Tess was also adapting well. She was still very much a loner and depended on Sophie much of the time. Although she enjoyed her own company, she would become upset if she didn't know where Sophie was. In fact she wouldn't go anywhere without her being close by. Unlike Naika, who never shared her food, Tess would often break off a branch full of berries and share them with her close friends Sophie, Naika, Bahati and Kiza. Because of her ravenous appetite, Tess was the most adventurous when it came to eating. She had also developed a taste for invertebrates, and was now eating ants and winged termites by the handful. Most of the others were still disgusted by these creepy-crawlies. Tess was also becoming quite proficient at making nests in the trees, and was the only one in her group to do this. Presumably she had gained some knowledge from her mother before being captured. She would spend ages perfecting her nest, which she would build in a bush at a height of about 12 feet. Then, when it was finished, she would lie in it for a short nap. Naika was also observing Tess's nest-building and would make her own attempts, although they weren't very impressive. Still, it was a start, and we were hoping that she would improve this skill over time.

Sophie was becoming a pivotal figure within the group. It wasn't long before Maisha had formed a close friendship with Sophie. This started one morning when the chimps were given sugar cane, but Sophie had had hers taken by Niyon. As there was no more sugar cane left, Sophie had to go without. When all the other chimps began eating, Sophie was understandably very upset. But Maisha broke off a piece of hers and gave it to her. Sophie was so grateful that she wouldn't stop kissing and panting at Maisha for ages. This was a good move on Maisha's part, as it cemented their friendship. From then on it was noticeable that Sophie would groom Niyon just before feeding time, as if this was her strategy to win his friendship, so that he didn't pinch her food again.

* * *

Inside the sanctuary the acacia trees were rapidly regenerating, and after four years we had yellow fever saplings growing in abundance. Elephant and giraffe were attracted to the lush growth and would spook the chimps, causing them to scramble to the very tops of the trees. Occasionally elephants would break into the enclosure by breaking a row of posts with their trunks, until the fence sagged to the ground, shorting the electric current. This was a major concern, and our worst nightmare was to have the entire group of adult chimps escape. Because the vegetation was thicker on the inside than on the outside and because each enclosure was over 100 acres (40 hectares) in size, there was no great incentive for the chimps to risk a shock from the fence in order to venture out.

However, we did have three instances when they decided to see if the grass was greener on the other side. Fortunately, the staff were well rehearsed on what to do. The essential thing was to remain calm, keep any tourists well away from the area and try to walk the chimp back inside. I had darting equipment in case things went wrong, but luckily I never had to use it. The staff were excellent in walking the chimps back to the sanctuary, especially Joseph Maiyo, who was the calmest and bravest of them all.

Despite the regular arrival of tourists, the chimp sanctuary was running at a loss, and the game reserve was barely breaking even. Tourists would pay $20 to visit the chimps, which included the boat ride along the river. Both sanctuary and reserve survived because of subsistence from the cattle on the ranch. Lonrho was under immense pressure to make both profitable. So I came up with a plan to let tourists visit the chimps free of charge and instead increase the gate price for the reserve by just 300 shillings — less than $5 — meaning that everyone visiting the game reserve would be contributing to the chimps. The idea was accepted by Lonrho. The game reserve became a more attractive proposition for tourists and overnight the chimp sanctuary broke even.

Meanwhile, with all the debates raging over whether to allow breeding, the chimps had their own views on this contentious issue, and Judy and Sultana became pregnant. One morning in early July 1996, Judy was seen vomiting, so as a precaution we separated her from the rest of the group. The following morning she gave birth to a baby male. This was now the second chimp I had seen vomiting prior to giving birth. Then the next day, at around seven in the evening, Sultana went into labour, and an hour later gave birth to a baby female.

We weren't sure who the fathers of these babies were but believed it to be one or both of the two younger males, Ndaronse or Gerbil. They had previously been kept in peer groups and had not endured prolonged isolation. As a result they were sexually active, unlike the other males, who rarely showed any interest in the females. Both these males had their favourite females; Ndaronse was especially fond of Alley, and Gerbil preferred Cheetah. When Cheetah came into oestrus we separated her from the group and kept her in a small outside enclosure.

Gerbil was terribly upset at being separated from Cheetah, especially as she was ovulating at the time. He moaned and groaned and rolled about on the ground in a tantrum. In the end he climbed over the electric fence, receiving many shocks in the process, just to be with her. We managed to separate him from her, only for him to climb back over the fence again. Ndaronse was close friends with Alley. A few times, when Alley was ovulating, Gerbil would manage a sneak mating. But if Ndaronse discovered them in the act, he would wait until they had finished and then rush over to Alley, bellowing in a fit of jealously and hit her several times across her face and body. It was interesting that he would never do anything to Gerbil, despite the fact that he was marginally dominant over him.

Meanwhile Judy was still separated from the others. We were now faced with the problem of reintroducing her and her baby to the rest of the group. This went fine until we introduced her to the dominant male, Safari. He ambled up to her, and was curious to inspect the baby. But Judy panicked and, inexplicably, threw her baby high into the air. It landed heavily on the floor in front of Safari, who seemed as surprised as we were by what had happened. He stood over the abandoned baby, and despite our efforts to distract him, bent down and bit it on the thigh. The baby screamed. Throughout this ordeal Judy made no attempt to intervene. But Sultana was distressed by the baby's screams and attacked Safari, grabbing Judy's baby in the process. For a few days she nursed both babies. Much to Judy's annoyance, she refused to give her baby back, and as a result Judy's relationship with Sultana deteriorated. But, sadly, Judy's baby died a few days later, presumably from its injuries.

I asked the staff to think of a name for Sultana's baby and later she was named Mwanzo, which means 'the beginning'. The death of Judy's baby was a shame for Mwanzo, because now she would have to live a life without another infant to play with. Mwanzo had an immediate impact on Sultana, and also on the dynamics of the group. Ever since Sultana's arrival at Sweetwaters, she had displayed a slightly neurotic demeanour. She lacked confidence and had problems dealing with the concrete floors in the building. She would always walk uneasily around the edge of the room, holding on to the walls, and would never walk across the middle of the floor. But having a baby changed all that. She now walked boldly across the floor unaided, and had grown in popularity, because all the members of her group wanted access to her baby. Sultana was initially very vigilant and suspicious of the adult males. The only male she allowed to interact with her baby was Ndaronse, who I suspected might be the father.

Another male chimp, Socrate, would lie flat on his back on the ground and try to reach out with his fingertips in an effort to play with Mwanzo. At other times he would do headstands to get the baby's attention. After a few weeks Sultana became more tolerant of others interacting with her baby, and allowed them to sit next to her. But she remained wary of Judy, who was constantly hovering close by. Judy was a disconsolate figure without her baby. She seemed very subdued and depressed. One morning, when Mwanzo wandered a short distance from her mother, Judy picked her up and hurriedly set off towards the bush. Sultana was immediately distressed and chased after her. Judy wisely allowed Sultana to take her baby back. It may well be that she believed Mwanzo was her own baby.

I was pleasantly surprised at how good a mother Sultana actually was. Over the next few months she encouraged her baby to crawl. She would place Mwanzo in front of her, and then take a couple of steps backwards, so that the baby had to crawl towards her. The baby clearly enriched the lives of every member of her group. Apart from Sultana, who developed self-confidence and maternal experience, the other females benefited indirectly by carrying Mwanzo from time to time. Tension was noticeably reduced among the males, as they loved playing with the baby.

* * *

Oliver was growing up fast, and was now almost five and a half. As wonderful and as educational as it was living in the African bush, he was missing out on a formal education with other children. We were also concerned that he was being deprived of daily contact with human peers and was starting to fall behind in school. Learning how to count baboons was not quite the same as learning arithmetic. So we decided that Audrey and Oliver would return to England for a school term, and arranged for him to join a local school outside Cambridge. They returned to England in August 1996 in time for the beginning of the new school year. This was all a big change for Oliver. His only real previous school had been the African bush. Now he was thrust into a life of ritual and discipline. Initially a little behind the others in his class, he soon caught up and formed his own group of close friends. Oliver was quite proud of the scar on his chest — after all, not every child has been bitten by a chimpanzee.

Back at Sweetwaters, Sophie was missing Oliver very much. She had spent most of her life with her human companion, playing with him and keeping a watchful eye on him. And she, of course, no longer had the individual attention and affection she had been receiving from her human family. Now she had to make do with her gang of hairy, uncouth friends. Sometimes a boat would take up to eight tourists at a time along the river as it slowly snaked its way through the sanctuary. The chimps would be waiting by the bank for the boat. Sophie always greeted the arrival of the tourists with enthusiastic hooting and arm waving. But if she heard or saw children she would stand on her two feet as tall as she could and peer into the boat to see if Oliver was among them. It was quite sad to see that she was obviously desperate to find him again.

Less than three months after Oliver started school, a producer from BBC TV's *Really Wild Show*, a popular children's wildlife programme, telephoned me in Kenya. She had heard about Oliver and Sophie, and wanted to film them together, as well as the other chimps, and capture their unique relationship. I explained that Oliver had just recently returned to school in England. Undeterred, the producer offered to fly Oliver and Audrey out to Kenya for filming.

This was a lucky break for us all and it came completely out of the blue. Oliver and the chimps were happily reunited again. The filming lasted three days, but Audrey and Oliver stayed a week in Kenya. The programme was presented by Howie Watkins, and he was filmed with Oliver in the sanctuary with all the chimps and then, over a couple of days at our house, with just Sophie, Tess and Naika. The short film was called *The Boy Who Talks to Monkeys*, which, understandably, really annoyed Audrey. But Oliver became an instant hero in his school, when he returned.

Meanwhile the debate was still continuing over the breeding policy. It was agreed at a steering committee meeting that we should place the chimps on the long-acting human contraceptive Norplant. This drug is administered as six matchstick-sized implants inserted into the upper arm. These give off very small amounts of a hormone that prevents the release of eggs from the ovaries, and are effective for up to five years. But by this time Judy had become pregnant again.

In December 1996 we finally arranged with Tom deMaar and Dr Butt, who was a friend of ours in Nanyuki, to perform the Norplant operations. Judy was left to continue her pregnancy and the remaining mature females were successfully placed on this contraceptive. Tom anaesthetised the chimps and Dr Butt inserted the Norplant rods.

Oliver and Audrey returned to Kenya in time for Christmas. He had completed his first term at school, and we decided to try to continue his education from home. He attended a part-time school in the neighbouring village of Timau, along with most of the other expatriate children from the region. Although the situation wasn't ideal, it was better than sending Oliver away to boarding school, which is what most white Kenyans did — something we could never contemplate.

In the middle of January several of the young chimps became ill with bad colds. Most of the group were affected, but Maisha seemed particularly ill. She had caught several colds during her short time at Sweetwaters and we were hoping that she would recover as usual. She didn't seem too bad, and was walking about with the other chimps during the day. However, the next morning she had lost her appetite and had a runny nose. We kept her inside, but by the evening her condition had rapidly deteriorated. Typically, she refused all medicine. I phoned the KWS vet, but there was no one available. They told me to try again in the morning. When I went to check on Maisha in the night, I found her in a bad way, gasping for breath. But it was too late, and she died shortly afterwards. We were all very depressed over her death. I had to call Mr Devani to break the news to him. He was devastated, just as Karl had been when Charlie died. It was ironic that Charlie and Maisha should both die and break the hearts of two men, and yet the chimps who had no previous owners — the unknown soldiers, so to speak — should be fit and healthy. I wondered if the same fate would befall me with Sophie.

Mr Devani arranged a funeral for Maisha. The death of Naika Devani was announced in the *Daily Nation* newspaper. I took her body to Ol Jogi Ranch for Tom deMaar to perform an autopsy. The result was inconclusive but showed that she had pneumonia. Mr Devani arranged for a coffin to be driven to Ol Jogi. He requested that we dress her in some children's clothes, which were delivered with the coffin. Although we felt uncomfortable with the idea, we respected his wishes and dressed Maisha in the white dress, white underpants and white gloves. Later that day she was buried in the family garden. Seeing her dressed in this human attire made her look somehow ridiculous and I felt it was a rather undignified ending for her — not the way I would personally want to remember her.

15

Kwa Heri Kenya

Life was becoming increasingly hard for everyone working at Ol Pejeta. Sadly, in order for wildlife conservation to work effectively it needs to be profitable. But, to me, Lonrho seemed to be in a financial mess. None of its business ventures appeared to be making money, and it wasn't long before Ol Pejeta Ranch was put up for sale. However, despite much interest, most of the offers didn't meet Lonrho's own valuation of the ranch. This left the management at Ol Pejeta looking over our shoulders, wondering if we would soon be out of a job.

Things went from bad to worse, when my boss, Russell Clarke, resigned from Lonrho and was replaced by Jim Taylor. It was David Heath who told me the news about Russell. I could see that David seemed agitated. He felt his own position within Lonrho was now under threat. He wasn't wrong and not long afterwards he was replaced by Richard Vine.

I had been at Sweetwaters for almost four years and was undecided about staying on any longer. But Lonrho informed me that they wouldn't be renewing my contract, which had six months left to run. They were quite good about it all, and at least allowed us to stay at Spoonbill

Dam for the duration of the contract. During this time there was a handover period and my successor, Ann Olivecrona, took over the running of the chimps. In a way I didn't have too many misgivings about leaving and felt I had achieved more or less what I'd set out to, especially with Sophie. I wanted to study for a PhD in Primatology and realised the time was right for me to move on. So we started to make plans for the future. We decided to return to England, get Oliver established in his school and then take things from there. A container was delivered to our house and we packed all our belongings into it. We shared a lot of our less valuable possessions with our three house staff. I promised them that I would find alternative jobs for them before I left. This proved harder than I imagined, but I eventually succeeded before leaving.

We threw a big party for all the staff, and drank and danced until dawn. A couple of days before leaving, I learned the sad news that Gerbil had died of pneumonia — an unfortunate start for my successor. It was a great shame, as Gerbil was a likeable rogue. Born in the Kibera Forest in Burundi, he was also the only chimp that wasn't from the Congo. But there was also some good news, as Judy gave birth to a healthy male baby, named Oscar. This brought a broad smile to my face. How nice to think that Mwanzo would have a friend her own age to grow up with. I was also pleased for Judy. Her maternal feelings had always been strong, and after all her suffering she was now, I hoped, about to live a fulfilled and contented life. This was, after all, the selective and limited breeding that had originally been agreed.

Although this was a sad time in our lives, we felt that we had achieved our aim as foster parents, and that Sophie would be safe and happy in Africa with her many chimp friends. She was now an important and influential member of a chimpanzee group and that left me with a sense of comfort and pride. Oliver took leaving Kenya in his stride. He was looking forward to going back to his school near Cambridge and meeting up with all his friends. It was great that Oliver enjoyed his school. I never had such a luxury and hated every day of my school years.

We drove to the chimp sanctuary for the last time, and spent the morning with Sophie, Tess and Naika. We were all very subdued. Sophie was always very perceptive of changes in behaviour and body language, but we made a good job of disguising our sadness and the chimps didn't notice anything different. The moment finally came when we said '*kwa heri*' to Sophie, the staff and all the friends we had made — both human and non-human. It was just three months short of Sophie's seventh birthday, and we each gave her a kiss and then got in our car and drove away for the last time.

We drove to Nairobi and booked into the Ambassador Hotel, as we wanted to spend a bit of time shopping for souvenirs in the city before catching our overnight flight to London. Later that evening I carried our suitcases to the car, placing them on the ground as I unlocked the boot. Stepping out from among the crowd, a man walked swiftly past, picked up one of our travel bags and coolly continued on his way. Audrey reacted instantly, shouting, 'Vince, someone's just stolen our bag!' Before I had time to do anything, around two dozen taxi drivers had formed a menacing lynch mob and chased the thief down the road. The man wisely dropped our bag and fled for his life. He had a lucky escape, for Kenyans don't tolerate thieves, and anyone who is caught stealing is usually lynched. This was a bit of a shock and a disappointing last memory of Kenya for Audrey to take back with her. But it would have been a disaster had we lost that particular bag, as our passports and plane tickets were inside.

Our luggage safe in the car, we drove to the airport. Along the Uhuru Highway I noticed something lying on the road. It was dark and there were no streetlights. I strained my eyes to see what it was. It looked like a sack on the road. As I approached, I realised that it was a man's body, and swerved at the last moment, narrowly missing him. As I looked back, I noticed that there was a police car parked further some way in front and two policemen were interrogating someone. I couldn't believe that they hadn't parked their car in front of the body to protect it or at the very least put some sign on the road to warn other drivers.

'Dad, what was that on the road?' Oliver asked. I told him it was nothing, probably just a sack of potatoes fallen from a lorry.

We flew back to England and stayed with Audrey's parents just outside Cambridge. After four years in Africa, Oliver had developed a slight Kenyan accent, which our family found amusing. We had decided that we didn't want to go back to Meadowtown. We needed a clean break and to start afresh, somewhere new. Plus, after four years in sunny Africa, we didn't think we could bear those Shropshire winters again. We managed to sell our house, and made a handsome profit on it too. Then we bought a smaller property just outside Cambridge, not far from Oliver's school. On his first day back the headmaster led him into his classroom and announced to all the children, 'Look who's back!' There was a loud, spontaneous cheer, and Oliver felt pleased to be back among friends. I am not sure if his early chimp social skills have been an asset to him but he has many friends. Like nearly all children, he is very much into computer games, a far cry from those days of mud wrestling with the chimps or riding on Tess's back like a young, proud Tarzan.

Audrey took a job as an administrator with a firm of accountants and I took a desk job with a company in Birmingham called Excell Multimedia while I explored various possibilities for studying for a PhD. Our three cats stayed in Kenya with David Vine. Sabby was now 14 and the males were both 13, and we felt it would be better for them to continue their lives without further disruption rather than spend the next six months in quarantine. Besides, we were planning on one day returning to work in Kenya.

However, not long afterwards we received the sad news that both Sabby and Tufty had died, within a few weeks of each other. Sabby had caught pneumonia during the rainy season and Tufty was believed to have been killed by a leopard. His head had been found in the bushes, but no body, one of the sure signs of a leopard kill. I suppose it was somehow inevitable that one of them would fall prey to some larger cat predator, and that Tinge, who had come through the incident with the lion unscathed, would be the one to survive. But at least we felt they had had a good life. Chapters in my life seemed to be closing all around me, and it was all rather sad.

After a few months I paid a visit to Chester Zoo. Neil was still working with the apes but Steve and Ross had both left to work for their local police force. The chimp group had now increased and numbered 28. They were breeding prolifically. The older ones instantly recognised me, especially Boris, who, on seeing me approaching in the distance, immediately began searching for something to throw at me. Then, having carefully selected a large wad of turf, he launched it across the moat — his way of saying hello. It was also a show of dominance, in case I had plans of returning to the fold. All the chimps were doing well, but I discovered that Gloria had died a year earlier, in 1996. Hearing about the death of a chimp you knew was like losing a friend. On a happier note, Mandy had given birth to yet another baby, Zee Zee, and was successfully rearing her. Maybe this time she would be more fortunate.

16

Life Begins at 40

Two years passed and I was still working in Birmingham. My plans to go back to university were constantly being frustrated. I had wanted to study chimp nest-building behaviour in Uganda's Budongo Forest, but found that there was already another student doing similar research. Then I tried to study bonobos in the Democratic Republic of Congo, but the political turmoil in the country after the overthrow of Mobutu made this too dangerous. Finally I was offered the chance to study chimps in Nigeria's largest National Park, Gashaka Gumti, but I was unsure about going to Nigeria, and took so long to make up my mind that another student took the place. Instead I was offered the alternative of studying baboons, which didn't really appeal to me.

We kept in regular contact with Sweetwaters by phone to get news of Sophie. But I was determined to go back and see her. So, after I'd booked a plane ticket, I telephoned Sweetwaters to alert them that I was returning for a visit. I was shocked to hear that only three days earlier, Steven, my former gardener, had suddenly been taken ill with cerebral malaria and pneumonia and had died. I was left feeling terribly deflated. I had been looking forward to seeing Steven again. He loved his job and the chimps, and I knew that Sophie would greatly miss him.

I left for Kenya on 1 July 1999; it was almost exactly two years since I had seen Sophie. After arriving at the airport I went immediately to find a taxi to take me to the city centre. The driver's attempts to rip me off made me feel instantly at home. I suppose that deep down, in a peculiar way, I had really missed all the bartering and haggling — something that previously infuriated me. I eventually agreed to a compromise on his *mzungu* price and we moved slowly out of the airport in his Nissan. Soon we were speeding towards the centre of Nairobi along the Uhuru Highway, which passes through the city's industrial suburbs. I could already see many changes in just two years. New factories and warehouses were cropping up along this road all the time.

To my right, a pair of crowned cranes were elegantly performing their courtship among a scattering of whistling thorn trees. As I observed the male dancing to his appreciative female, a couple of young Masai warriors came into view on my left, herding 20 or so Boran cattle through the scrubland. We were barely five minutes out of the airport and I was already witnessing wildlife and Masai culture, two symbols of Kenya, striving to coexist, ironically juxtaposed with modern industrialisation. Twenty minutes later we entered the city. I had forgotten just how alive Nairobi really is; bustling, vibrant and throbbing with noise and activity wherever you looked. As we reached the centre, the traffic came to a standstill and drivers became more animated, gesticulating angrily at each other. There seemed to be more cars than ever before queuing along the roads, honking their horns, like migrating wildebeest waiting to cross the Mara River, and spewing their clouds of thick black diesel fumes from their rears like wildebeest kicking up the Mara dust. Even the streets seemed more crowded with people. You had to wonder just where so many people could all be going: were they travelling to and from work or were they just strolling aimlessly and gossiping on street corners to pass the time of day?

I felt instantly at home, as if back at last in my home city and among my own people. I booked into a hotel for the night and then decided to wander around Nairobi. My hotel was in Westlands, a few miles outside the city centre, so I searched for a *matatu* (a small local bus) to take me to the centre of town. Eventually I found a half-empty one and sat at the back. The passengers waited for the *matatu* to fill up so that we could begin our journey. Touts were furiously trying to tempt people to take a ride rather than walk. In the end they needed just one more person for the bus to be full. Everyone was now becoming impatient for the journey to begin, and were watching the touts desperately trying to fill the last empty seat. In Nairobi there is humour to be found in everyday life, and the touts were like comedians holding centre stage. They asked one young woman if she wanted a ride and she flatly refused. One of them then grabbed her by the arm and began pulling her towards the vehicle. She was laughing and insisting that she wanted to continue on foot. In the end she managed to break free of his grasp. The next victim was an elderly man. This time both touts grabbed him and forced him inside. The man kept insisting he didn't want a ride, but they were determined, and forced him into the one remaining seat, then closed the door. The *matatu* drove off, leaving everyone falling about with laughter. Even the old man could see the funny side.

I stopped for lunch at the Ambassador Hotel, my old stamping ground. A waiter brought me a pot of tea and a jug of hot milk. One of the ways to tell a tourist in Africa is to watch them pour milk into a cup. Milk jugs in Kenya are designed to annoy you, and if you don't have the knack, then the milk will go everywhere but in the cup. So I poured mine and watched as it spilled down my wrist. It was obvious that I had been away too long!

That evening I had a pizza at one of the Italian restaurants in Westlands. As I was walking back to the hotel, I heard footsteps running towards me. I turned around, just as two men rushed up and grabbed me by my arms. They showed me their IDs and said they were police. One was much older than the other, with terribly protruding teeth. They asked me what I was doing out at this time of night. I said it was only nine in the evening and I was on my way back to my hotel. They asked to see my passport. Typically, I had left it in my suitcase in my room. I told them that I wasn't carrying it with me for fear of having it stolen. This was partly true anyway.

'How do we know that you are not selling drugs?' said the man with the bad teeth. 'We will have to arrest you and you can stay the night in a cell,' said the younger man.

'Look, my hotel is just around the corner, only a few hundred yards away. If you come with me, I'll show you my passport,' I told them.

But they weren't interested in being considerate. They were aggressive and intent on trying to intimidate me. But I remained calm.

Then the one with the bad teeth said, 'You know, if we go to your hotel room, we *will* find drugs there.' It was obviously a veiled threat that they were going to plant something in my room. Clearly what they wanted was a bribe. Luckily, I happened to have an old Lonrho business card on me from my days at Sweetwaters.

'If you take me to the police station, then I will have to telephone Mark Too,' I said, showing them my business card. Mark Too is the Chairman of Lonrho and very close to President Moi. Their jaws dropped, and they immediately told me that I could go but next time to make sure that I had my passport with me.

The following morning I was picked up at my hotel by Richard Vine's driver and then driven to Nanyuki and straight to Ol Pejeta. As we drove through the ranch, our vehicle kicked up clouds of dust, and it felt great to be back. I popped in at Richard's house to see how Tinge, our one surviving cat, was getting on. Karanja, our former cook, who was now working for Richard, found Tinge for me, sleeping on the roof of the kitchen. He had grown fat in those two years and looked older than his 14 years, with his fur all dusty and tatty. Then I was invited for a meal at David Bear's house, where I also stayed the night. Dave had replaced Ian Agagliate as Ol Pejeta's mechanic and workshop foreman.

The next morning Dave drove me down to the game reserve, where they were about to release a young male leopard. This handsome beast had been killing sheep on a nearby ranch. So the KWS had no alternative but to capture him. Sweetwaters had agreed to give him a last chance of survival in the reserve. I stood back and watched as Dave released the leopard, which leapt from the trap and scampered off into the bushes, probably never to be seen again. The task completed, we got back into the car and made our way towards the chimps.

We rounded a bend in the narrow dirt road and the chimp building came into view, some 50 yards away. My heart started beating faster with anticipation. I told Dave to stop his car here so that I could prepare my video camera. I then walked the rest of the way towards the chimps, who were all waiting for their food. As I appeared, they all gave out loud hoots. Somewhere among the din I could recognise Sophie's voice screeching, but I couldn't see her. I said my hellos to all the staff. It was nice to see them again and I was pleased that they were still dedicated to the chimps. I asked where Sophie was and they told me that she was on the other side of the building in a small enclosure because she was in oestrus. Sophie could see me through the fence in the distance. I walked around the building and found her with Tess, waiting for me by the fence. Sophie gave off a rapid pant greeting, and I reciprocated.

It was immediately obvious to me that she had grown up in the two years. She was showing a half swelling of her genitals and was fast becoming sexually mature, although not yet mature enough to conceive. Adult female chimps exhibit a pink swelling of their genitals when they are in oestrus. This signals to the males that they are sexually receptive. Gone was the frantic screaming that I expected. Instead she seemed much calmer. I was amazed at how big Tess had grown. Her face had become darker and she looked so much older. In fact, other than by her fat tummy, I could hardly recognise her.

As I chatted with the staff, they told me how the chimps were all getting on: Sophie and Naika were apparently very popular with the others, but Tess was still a bit of a loner. Naika hadn't changed at all and was still as boisterous as ever. I learned that, apart from Sophie, all the chimps had become coprophagic — eating their own excrement. Chimps in the wild are only slightly coprophagic compared with, for example, gorillas, who habitually eat their own faeces. They obtain some nutritional value from undigested food remains, such as nuts and seeds. In captivity, primates often perform this action out of boredom or as a result of an inadequate diet. But I wasn't surprised that Sophie didn't eat her faeces, as she had always been so careful not to touch any. Dickson then told me that there were plans to place the other adolescent females on the Norplant contraceptive.

The staff allowed me inside with Sophie and Tess. Sophie immediately jumped into my arms and gave me a tight hug with a long, hot, panting, open-mouthed kiss on my neck. I did the same to her and we panted in harmony into each other's necks for a few minutes. Greetings over, she swung around my body and in the same motion positioned herself on me piggy-back style. She had grown much heavier since I had last carried her. We walked around for a while like this, then sat down together for a long grooming session.

As ever, Sophie was fascinated by my shoelaces and spent a lot of time undoing them and re-threading them. I suppose for her they were like a puzzle is to a child. Everything about her seemed more grown-up and mature, and it made me feel like a parent visiting a teenage daughter. But it was comforting to see her as a young adult. She appeared contented, more independent and more confident. We stayed together for several hours and then it was time for me to go, as I was relying on Dave to take me back to the ranch. Because this was only going to be a short trip, I felt easier about having to disappoint Sophie by disappearing again.

I returned again the next day and spent the morning with her. We played one of her favourite games inside the building — a game of catch. Sophie would push her hand through the bars and wait for me to try to grab it — as she had done as a baby on the bed with Oliver all those years ago. I was anxious about spending too much time with her in case she became upset when I left. She was fully weaned off me now, so I didn't want my visit to be a backward step.

I crossed over the river and went to see the adult chimps. The staff called the chimps over to me and gave them sugar cane. Sultana and Judy arrived together, carrying their babies on their backs. It seemed that Judy had finally forgiven Sultana. Sultana settled down in the shade of a bush and bit off a large chunk of sugar cane and then held the remainder in her lap so that Mwanzo could get her share. Mwanzo and Oscar had grown into large, healthy infants. They wrestled all the time and seemed great friends. I couldn't help thinking how fortunate these two babies were. But my enjoyment was later spoilt by the news that the female Furaha had recently died after being bitten by a venomous snake.

Joseph Maiyo then told me a horror story of how a ranger had come close to being killed by one of the chimps. Chimps have long memories and definitely bear grudges. Their reaction to anyone carrying a gun was a clear indication that they hadn't forgotten the day their mother was slain by a human armed with a shotgun. They knew that a gun was dangerous and had witnessed what it could do, and if they saw anyone carrying one they would all instantly become terrified and aggressive.

Every day tourists would visit the chimps. Normally a ranger would accompany them and take them to a viewing platform where they could watch the adults from an elevated position. Then, if they wanted, they could go on a short boat journey to watch the infant chimps by the river. One particular day the ranger, Samuel Kipchumba, had brought a group of tourists to the viewing platform. But Samuel was carrying a gun — something Poco hated with a vengeance. Poco began displaying aggressively, running up and down and hitting the fence with a stick. Suddenly he shot his arm through the electric fence and grabbed hold of Samuel. Ignoring the shock he got from the fence, he attempted to pull Samuel towards him. Joseph was standing nearby and grabbed hold of Samuel's arm, preventing him from being pulled inside. But as he yanked Samuel away from the fence, Poco was also dragged through a gap. Now he was outside and intent on revenge. He continued his relentless attack on his poor victim, biting him many times on the legs, arms and hands. In fact he virtually bit Samuel's knee-cap off.

Richard Mutai rushed to help Joseph and the two of them began hitting Poco with branches from a nearby bush. Poco was in such a frenzy that this had little effect. Luckily, quick-thinking Joseph, seeing Samuel's gun lying on the ground, grabbed it and hit Poco on the head with the butt, knocking him unconscious. They lifted Samuel up and carefully carried him back to the building, where they placed him on the back of a pick-up. But Poco regained consciousness and raced towards the car. Fortunately, the driver managed to get away in the nick of time, and took Samuel straight to the hospital, where he made a remarkable recovery. The staff then successfully walked Poco back into the building. Joseph and Richard had risked their lives to save their friend — an act of heroism that undoubtedly saved Samuel from a horrific death.

I thanked all the staff and told them I would return for a longer visit next year. I said goodbye to sweet Sophie and then I was driven to Nanyuki, where I caught a taxi back to Nairobi. As we drove back to the big city, I thought about Sophie and Sweetwaters. I could see that Sophie's and Oliver's parallel lives were rapidly diverging as they grew up in their separate worlds. They had developed physically at more or less the same pace throughout their young lives, but now Sophie's development had accelerated. She was rapidly becoming a young adult while Oliver was still very much a child.

⋆ ⋆ ⋆

I returned to England with many photos and a video of Sophie and the rest of the chimps. Audrey and Oliver were very excited to hear about my travels. That evening we settled down in front of the TV with dinner on our laps and watched my video of Sophie and the other chimps.

A week later I arranged a meeting with Chester Zoo's director, Dr Gordon McGregor-Reid. Gordon had replaced Michael Bramble, who had retired. I let him know how Sophie was doing and explained that I was concerned that she wouldn't be allowed to breed and that she might be put on a contraceptive before she was even a fully developed adult. I asked him if the zoo would consider giving me joint ownership of Sophie with the zoo. That way I would be able to keep a close eye on her welfare and development, and ensure that nothing could happen to her without my being informed.

It was important for me that Sophie should be allowed to breed. I knew from her maternal reaction towards Annie, the vervet monkey, that she and the rest of the chimps in her group would benefit enormously from the opportunity to express and develop maternal skills. That was, after all, the CITES 'breeding loan' agreement that allowed her to go to Kenya in the first place. I would then feel that I had achieved an important objective as a foster parent.

Gordon sounded sympathetic, and said the zoo would consider my request. However, after about a month, I received a letter from him stating that after careful reflection they had decided not to agree to my wishes. I was disappointed and still believe it was the wrong decision. However, I accepted their judgement. But I wasn't going to give up so easily, and decided to let things lie for the present, while holding on to the idea of trying again some time in the near future.

But Sophie was constantly preying on my mind. I wanted to be able to sleep peacefully at night knowing that I had given it my best shot and exhausted all possible avenues. So I phoned Jane Goodall at her home in England. I told her that the chimps were doing fine, but that I was concerned for Sophie's welfare, and anxious that the KWS might be about to put her on a contraceptive or, even worse, sterilise her. Jane assured me that there had been no decision to put any of the infants on contraceptives. I wasn't sure if she was just saying this to make me feel better but it left me a little more reassured. Jane also told me that the old male chimp Grégoire had been successfully moved from Brazzaville Zoo and was now living happily in the sanctuary at M'Pili.

* * *

On 11 November 1999 I awoke from a heavy night's sleep. Today was going to be a special day: my fortieth birthday. Audrey had made plans to spoil me and there was going to be a party at our house in the evening with family and a few friends. In the morning she took Oliver to school and then continued on her journey to work. So I had a lazy morning and got up late, at around ten o'clock. I dragged myself out of bed and went downstairs to make a cup of tea. I wasn't too depressed at the thought of turning the big four-zero, as I had already grown resigned to this over the previous months. So I decided to indulge myself with a healthy fry-up of sausages, bacon, eggs and mushrooms.

As I was putting some oil in the frying pan, the phone rang. It was Richard Vine, the general manager of Sweetwaters. I was surprised to hear his voice and even more surprised that he should remember my birthday. But the nervous tone of his voice soon made me realise that this wasn't why he was phoning.

'Vince, I've got some very bad news to tell you.'

'Oh my God! Please don't tell me something's happened to Sophie. She hasn't died, has she?'

I don't know why I immediately thought the worst. But I couldn't imagine anyone phoning me from Kenya unless it was very serious. I was clinging to the desperate hope that he would say no and that maybe she was just ill or something. Instead he replied that he was very sorry but there had been a terrible accident and Sophie had died. My legs buckled under me and I cried out, 'Oh no! Oh no!' I was completely destroyed.

'Richard, today is my fortieth birthday. It's supposed to be a special day for me,' I said.

He again apologised and asked me if was there anything he could do. I begged him not to let anyone do a post-mortem on Sophie but to bury her near the chimp house in a marked grave. He agreed to do this and after apologising again, left me to my moment of despair.

I immediately phoned Dr Butt, our friend in Nanyuki, and asked him if it was true that Sophie was dead. He confirmed that it was, and that he had been there when it happened. I asked if there was any way that they could have made a mistake, that she was lying in a deep coma or something. I suppose in my moment of desperation I was clutching at straws. He replied that this was impossible, and then told me his version of what had happened. This was backed up by the official medical report, which was sent to me later.

The steering committee, he said, had decided to put Sophie on Norplant, even though she probably had another year before she would be able to conceive. The four vets who were taking part tried to dart Sophie but, being her usual determined self, she wasn't going to give in without a fight. They darted her twice with ketamine, but each time with no evident effect. Either she was 'kicking' the effects, or else none — or not all — of the anaesthetic had entered her. After a while they gave up and tried to calm her down by giving her a sedative; they gave her two tablets, but again with little effect. They gave her another two tablets. Although still adolescent, she had apparently now been given the dose for an adult human. According to Dr Butt, she was given a total of seven tablets, each one with juice or banana (a serious risk during any anaesthesia).

Sophie collapsed on the ground. Her semi-conscious body was then injected with ketamine, whereupon she suffered a total respiratory collapse. Attempts to perform a tracheotomy, and even mouth-to-mouth resuscitation, proved futile. According to Dr Butt, the whole affair was a total amateurish shambles. So appalled was he by the episode, he said, he would never assist in any operations at Sweetwaters again.

I put the phone down and dragged myself upstairs, where I collapsed on the bed and burst into tears. This had to be a bad nightmare — soon I would wake up. But I didn't wake up, and it wasn't a dream.

What made Sophie's death all the harder to bear was that she didn't die of natural causes. She didn't die of some disease, fall out of a tree, drown in the river or get killed by a wild animal. She died pointlessly and needlessly at the hand of man. Why was a chimp on a CITES breeding loan being put on a contraceptive anyway? After all, there was no danger of her becoming pregnant for at least another year. If only Gordon McGregor-Reid had said yes to the Sophie ownership issue, I would have stopped them. Had I been there, I would have never allowed them to treat any of the chimps in this way, let alone Sophie. I would have insisted that after the first failed attempt, they try again on another day, when she had calmed down, or use a different strategy. To top it all, Sophie could easily have been injected by hand through the bars.

I knew that I must break the news to Audrey. I phoned her at work. She could scarcely believe what I was saying. We discussed whether or not to tell Oliver, and decided we had no choice. When he heard what had happened, he initially went very quiet, then he ran to his room and cried — although I think it was more from seeing his dad so upset — something he had never witnessed before. My fortieth birthday party was cancelled.

* * *

Throughout that night, my tears flowed, soaking my pillow. I lay quietly on my back staring up at the dark, trying as best I could not to wake Audrey or Oliver and upset them any more than events already had. It was the worst night of my life. And there was no relief when morning came. For several days I remained inconsolable.

They say life begins at forty. Well, in my case it didn't. A great part of me died for ever on the morning of that 'special' day of mine. Of course, I wish that I could have written a happier ending, but sadly real life isn't always so kind. I hope that in some way Sophie's story will provide some insight into the world of our nearest living relatives. During her short life she gave much happiness to the few humans who had the good fortune to enter her world, and especially those who were proud to have been chosen by her as her friends.

Other titles published by
The House of Ulverscroft:

TICK BITE FEVER

David Bennun

Africa has a rich history of heroes and trailblazers. David Bennum is neither. He belongs to the dark continent's less celebrated tradition of accidental adventures. Throughout his 1970s African childhood, and despite the best efforts of his long-suffering family, David manages to get himself in more trouble than one person has any right to survive. Here is the story of how a bemused and clumsy small boy discovered Africa. A day rarely passes when he doesn't run the risk of getting eaten, crushed, poisoned, drowned, trampled, shot or impaled. Even his dog, Achilles, seems to have a death wish. David Bennun's writing is evocative, touching and so funny that you will find it hard not to laugh out loud.

RELATIVE STRANGERS

Hunter Davies

On 18 May 1932, Kate Hodder gave birth to a baby girl at Birchfield House Infirmary near Sevenoaks in Kent. Three and a half hours later, a baby boy was born. And then, forty minutes on, a second girl was delivered. But the joy at the birth of triplets — a very rare event in the 1930s — was to prove short-lived. Kate died the following day. Her husband Wills was soon struggling to cope with their six previous children, so the decision was made to offer the triplets up for adoption. Remarkably, aged sixty-nine, the three triplets were reunited. Here, their astonishing story charts their three different upbringings, and the struggles and detective work that brought them back together.

THE STORYTELLER'S DAUGHTER

Saira Shah

Saira Shah grew up in Britain, but she was always told she came from somewhere else: a fairytale land of orchards and gardens. The country was Afghanistan, the storyteller her father. Then, aged twenty-one, Saira set out to find the truth about her family's homeland. Instead of finding a paradise, she was plunged into a country at war. The journey spanned more than fifteen years. Whether extricating herself from an arranged marriage, walking through minefields with the mujahidin, or slipping clandestinely into the Taliban's Kabul, Saira learnt the bitter limits of the stories she loved. But she discovered the reality of a country more complex and challenging than anything she could have imagined.

THE CLOUD GARDEN

Tom Hart Dyke & Paul Winder

A place of legend, the Darien Gap is an almost impregnable strip of swamp, jungle and cloud forest between North and South America. Stories of abduction and murder there are rife. In 2000, young botanist Tom Hart Dyke set off to Central America to search for rare orchids. Pure chance brought him together with another young explorer, Paul Winder, in northern Mexico. Ignoring a warning from the LONELY PLANET guide — 'Don't even think about it!' — Tom and Paul set off into the Darien. For six days they made good progress. Then, near Colombia, they were ambushed by FARC guerrillas, who were to hold them hostage for the next nine months. Their survival was a matter of extraordinary endurance, incredible ingenuity and not a little good luck . . .

EMMA'S WAR

Deborah Scroggins

Emma McCune's passion for Africa, her unstinting commitment to the children of the Sudan, and her striking beauty and glamour set her apart from other aid workers from the moment she arrived in southern Sudan. But no one was prepared for her decision to marry a local warlord — a man who seemed to embody everything she was working against — and to throw herself into his violent quest to take over southern Sudan's rebel movement. Deborah Scroggins — who met McCune in the Sudan — charts the process by which McCune's romantic delusions led to her descent into the hell of Africa's longest running civil war.

TOMORROW TO BE BRAVE

Susan Travers

This is the story of Susan Travers' extraordinary life, from her childhood in England, her girlhood in inter-war Europe, her decision to join the Free French in search of adventure, her part in the North African campaign and, most remarkable of all, her time after the war in the Foreign Legion as a regular serving officer — the only woman ever to have achieved this. It is a tale of exceptional courage against overwhelming odds, and a passionate love story played out against the epic landscape of the desert, as Susan prepared to risk everything for the country and for the man she loved.